CHEMICAL ENGINEERING METHODS AND TECHNOLOGY SERIES

ENVIRONMENTALLY HARMONIOUS CHEMISTRY FOR THE 21ST CENTURY

CHEMICAL ENGINEERING METHODS AND TECHNOLOGY SERIES

Treatment of Tannery Effluents by Membrane Separation Technology
Sirshendu De, Chandan Das, and Sunando DasGupta
2009. ISBN: 978-1-60741-836-8

Handbook of Hydrogels: Properties, Preparation & Applications
David B. Stein (Editor)
2009. ISBN: 978-1-60741-702-6

Handbook of Hydrogels: Properties, Preparation & Applications
David B. Stein (Editor)
2009. ISBN: 978-1-61668-167-8 (Online Book)

Non-Ionic Surfactants
Pierce L. Wendt and Demario S. Hoysted (Editors)
2009. ISBN: 978-1-60741-434-6

Handbook of Membrane Research: Properties, Performance and Applications
Stephan V. Gorley (Editor)
2010. ISBN: 978-1-60741-638-8

Chromatography: Types, Techniques and Methods
Toma J. Quintin (Editor)
2010. ISBN: 978-1-60876-316-0

Green Composites: Properties, Design and Life Cycle Assessment
François Willems and Pieter Moens (Editors)
2010. ISBN: 978-1-60741-301-1

Environmentally Harmonious Chemistry for the 21st Century
Masakazu Anpo and K. Mizuno (Editors)
2010. ISBN: 978-1-60876-428-0

CHEMICAL ENGINEERING METHODS AND TECHNOLOGY SERIES

ENVIRONMENTALLY HARMONIOUS CHEMISTRY FOR THE 21ST CENTURY

MASAKAZU ANPO
AND
K. MIZUNO
EDITORS

Nova Science Publishers, Inc.
New York

For permission to use material from this book please contact us:
Telephone 631-231-7269; Fax 631-231-8175
Web Site: http://www.novapublishers.com

LIBRARY OF CONGRESS CATALOGING-IN-PUBLICATION DATA

Environmentally harmonious chemistry for the 21st century / editors, Masakazu Anpo and K. Mizuno.

p. cm.

Includes index.

ISBN 978-1-60876-428-0 (hardcover)

1. Environmental chemistry. 2. Environmental chemistry--Industrial applications. I. Anpo, Masakazu. II. Mizuno, K.

TP155.2.E58.E68 2009

660.028'6--dc22 2009036060

Published by Nova Science Publishers, Inc. ✦ New York

CONTENTS

Preface **vii**

Chapter 1 Bioanalytical Systems with *In Vivo* Biosensors and Microfluidic Devices **1**
T. Yao and H. Hisamoto

Chapter 2 Development and Application of Functional Inorganic Materials **33**
Masahiro Tatsumisago

Chapter 3 Development of Environmentally-Friendly Artificial Photosynthetic Catalysts Using Clean and Abundant Solar Energy **49**
Masaya Matsuoka and Masakazu Anpo

Chapter 4 Eco-Friendly Electrochemical Devices Using "Water" and "Hydrogen" as Key Substances in the 21st Century **75**
Hiroshi Inoue

Chapter 5 Environmentally Harmonious Organic Photochemical Reactions **89**
Kazuhiko Mizuno, Hiroshi Ikeda and Hajime Maeda

Chapter 6 Creation of New Organic Molecules, New Reactions Based on Heteroatom Chemistry **123**
Akiya Ogawa, Akihiro Nomoto and Motohiro Sonoda

Chapter 7 Utilization of Light in Polymer Science **137**
Masamitsu Shirai

Chapter 8 Polymer-Based Nanomaterials for Drug Delivery **149**
Kenji Kono

Chapter 9 An Environmentally-Friendly Plastic Coating Technique Using Gold Nanoparticles for Nanoelectronic Device Applications **163**
Tsutomu Nagaoka and Hiroshi Shiigi

Chapter 10 Polysilanes: Novel Sigma-Conjugated Polymers **181**
Kunio Oka

Index **205**

PREFACE

Industrialization of the ammonia synthesis system with the Fe-based catalyst developed by Bosh Harbor in 1890 has led to various important production systems using fossil fuels and the subsequent growth of the manufacturing sector in all parts of the world. As a result, we have inevitably suffered great environmental destruction and pollution such as the greenhouse effect, the presence of various dangerous and toxic substances in our surroundings, water and food, and serious climatic changes that are only now beginning to capture our attention.

This unprecedented explosion of modern industrialization in the 20th century has now led us to seek the development of new, sustainable and environmentally–friendly chemical processes which are safe and harmonious with our environment and can yet sustain our prosperity. The Department of Applied Chemistry of Osaka Prefecture University has initiated research and academic studies on such environmentally-harmonious chemistry strongly desired for life in the 21st century and beyond, publishing the book, "Environmentally-friendly Chemistry" in Japanese in 2000. This book has contributed to raising awareness of these issues with its high circulation and the professors of our university have used this book in their lectures related to the environment and/or sustainable chemistry. Now, with a desire to reach the international community, we would like to publish this book on this subject in English.

This new book consists of 10 Chapters and each chapter introduces the concept of environmentally-friendly chemistry from various fields such as Analytical Chemistry, Inorganic Chemistry, Physical Chemistry, Electrochemistry, Organic Chemistry, Organic Yynthesis Chemistry, Polymer Chemistry, Polymer Drug Delivery Chemistry, Sensing Chemistry, and Polysilane Chemistry. University students as well as graduate students at every level and in any scientific field may find it quite interesting and useful.

We sincerely hope not only that it arouses interest in sustainable chemical processes and systems but also that our work may contribute to the protection and safeguarding of our precious environment for future generations. Thank you very much.

Editors

Masakazu Anpo
Kazuhiko Mizuno

Osaka Prefecture University

In: Environmentally Harmonious Chemistry… ISBN: 978-1-60876-428-0
Editors: M. Anpo and K. Mizuno

Chapter 1

BIOANALYTICAL SYSTEMS WITH *IN VIVO* BIOSENSORS AND MICROFLUIDIC DEVICES

T. Yao and H. Hisamoto

ABSTRACT

The *in vivo* monitoring of the changes in concentration of biomolecules at a variety of sites for living systems including the blood stream, subcutaneous tissue, muscle, internal organs, and brain, has been desired for pharmacological, clinical, and biological studies. The most general approach to achieve such goals has been the use of implanted biosensors, however, it was limited due to variations in sensitivity during prolonged *in vivo* monitoring. Recently, microdialysis has been used as a sampling technique for *in vivo* analysis. This chapter reviews novel micro-flow injection biosensor systems which make possible the on-line monitoring of glucose metabolism and the pharmacological behavior of neurotransmitters in the extracellular space of the brain. This approach was developed by coupling a microdialysate sampling to a biosensor system using a flow-injection interface.

On the other hand, microfluidic devices fabricated on glass or plastic microchips have become one of the largest active research fields in recent years. By using the specific characteristics provided by the liquid microspace-like microchannel, attractive analytical performance features such as a fast response time and small sample volume are easily obtained. Another interesting feature of microfluidic devices is the formation of a multilayer flow which can be formed when the organic and aqueous solutions are made in contact with each other inside the microchannel. Employing the characteristics provided by the liquid microspace and multilayer flow allows us to develop novel analytical methodologies which cannot be achieved by other existing methodologies. Here, a multi-ion sensing device and biosensing device based on an inner-channel membrane immobilizing enzyme will be presented as examples.

Advances in these research areas are reviewed along with the results obtained in our studies.

1.1. INTRODUCTION

Microdialysis is a sampling technique for *in vivo* analysis [1] and has successfully been applied in pharmacological and biological studies of biomolecules in dialysate. In many cases, however, the dialysate was collected and usually assayed by liquid chromatography. The use of miniaturized biosensors was the most simple approach for *in vivo* monitoring, however, it was limited because of the variations in sensitivity during prolonged *in vivo* monitoring. In recent years, sensor systems combined with microdialysis sampling have been developed for *in vivo* analysis and the continuous monitoring of changes in concentration of biomolecules in blood, subcutaneous tissue, and the brain. For such research, some miniaturized enzyme-based amperometric biosensors have been proposed as specific detectors for the *in vivo* monitoring of glucose [2, 3-5], L-glutamate [6-8], acetylcholine [9], and L-lactate [2]. A similar attempt was also made by using an enzyme reactor as a recognition element for a particular analyte in dialysate, e.g., glucose [10]. Glucose is a main nutrient in the brain, while L-lactate and pyruvate are formed from glucose in an actively glycolyzing system, and are of interest in metabolic disorders. Also, L-glutamate and acetylcholine are the most important excitatory neurotransmitters in the brain. Therefore, *in vivo* monitoring of concentration changes in such analytes that take place rapidly in brain cells has been especially desired in the research field of brain chemistry.

On the other hand, microfluidic devices have also attracted much attention in recent years [11, 12]. Historically, this research field has progressed together with micromachining techniques such as photolithography patterning on silicone or glass substrates. Since the pioneering work of Manz et al., a new concept, the so-called micro total analysis systems (Micro-TAS) [13], has been widely accepted not only in analytical chemistry but also in systems biology, synthetic chemistry, chemical engineering and other fields. Among the many possible applications, microfluidic-based analytical devices are the most promising in the field of analytical chemistry due to the importance of the small sample/reagent solution volume or multiple analyte sensing which are among the most classical problems in this field. Recent advances in genomics, proteomics, metabolomics, all of which require multiple analyte determination in a very small sample volume, have made this a significant area of research.

The first half of this contributed chapter reviews *in vivo* micro-flow injection biosensor systems with an on-line microdialysis sampling which make possible the specific detection of a single analyte or the simultaneous monitoring of plural analytes in the brain. This approach has been developed by coupling a microdialysis sampling to the biosensor system using a flow-injection interface. The latter half reviews the general aspects of developing microfluidic-based analytical systems involving the characteristic features provided by liquid microspace and multilayer flow formation. Various applications using characteristics specific to a microfluidic system, such as ion-selective microextraction-based multiple ion sensing and enzyme-immobilized inner-channel membrane-based biosensing, will be presented.

1.2. Principle of Bioanalytical Sensors

Bioanalytical sensors (biosensors) are based on the direct combination of a biological receptor and a signal transducer with an electronic amplifier (figure 1.1). The effects of the interaction between a biological receptor and analyte are quantified as electrical output signals by the appropriate transducer and electronic part of the biosensor. Biosensors are classified into two types, namely, bioaffinity sensors and biocatalytic sensors (table 1). In bioaffinity sensors, dyes, lectines, antibodies, hormone receptors, or nucleic acids are used as molecular recognition elements in the construction of sensors for proteins, saccharides (glycoproteins), antigenes, hormones, and DNA. The physicochemical changes caused by complex formation are quantified by the appropriate transducers, such as optoelectronic sensors, electrochemical sensors, or thermistors. Another type of biosensor is termed biocatalytic sensors. In this type, enzymes, organellas, microbes, or tissue slices are used as molecular recognition elements for the detection of substrates, cofactors, inhibitors (activators), or also enzyme activities. By enzyme catalyzed reactions, analytes are converted chemically to various respective products. Amperometric and potentiometric electrodes, fiber optics and thermistors are the preferred transducers. As a general rule, the immobilization of an enzyme or antibody provides a preparation that combines high selectivity with increased stability. In addition, if the immobilized enzyme or antibody is used in a micro-flow system or microchip, handling is minimized and reproducibility is enhanced. Such flow-through biosensor systems are generally classified into system A and system B by the differences in the configuration of the immobilized biomolecules. In system A, a biomolecule is immobilized on a transducer in the form of a membrane. In system B, a biomolecule immobilized onto an insoluble support is packed into a small column which is positioned in the upstream of a detector. In this chapter, the authors will describe the development and applications of such flow-injection biosensor systems for *in vivo* analysis as well as microchips for bioanalysis.

Table 1. Receptors and transducers in biosensors

Bioaffinity sensors	Biocatalytic sensors
Receptor	Receptor
dye	enzyme
lectin	organelle
enzyme	microbe
antibody	tissue slice
receptor	
nucleic acid	
Transducer	
Potentiometric, amperometric, and impedimetric electrodes; optical detector; piezoelectric crystals; thermistors; field effect transistors.	

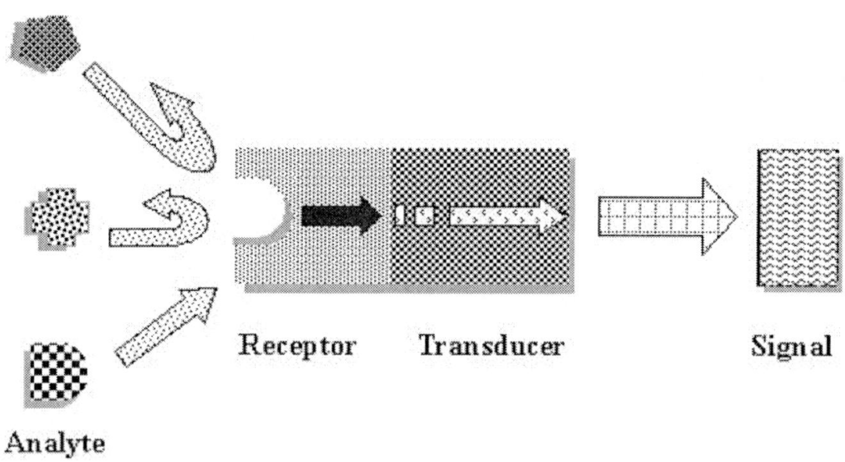

Figure 1.1. Configuration of a biosensor.

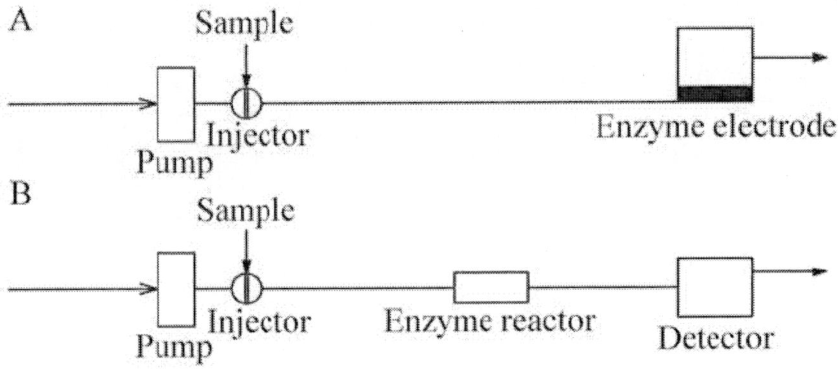

Scheme 1.

1.3. *In Vivo* Analysis of Biomolecules in Brain,7"

1.3.1. The Use of the Enzyme Reactor as a Recognition Element

1.3.1.1. On-Line Amperometric Assay of Glucose, L-Glutamate, and Acetylcholine Using Immobilized Enzyme Reactors [14]

Here, we describe an on-line amperometric monitoring system of glucose, L-glutamate, and acetylcholine. These analytes are not detectable by direct electrochemical oxidation. Therefore, we chose to use glucose oxidase (GOD), L-glutamate oxidase (GlOD), and acetylcholine esterase (AcE) + choline oxidase (ChO), to convert each of these analytes to hydrogen peroxide which can be detected amperometrically, according to the following equations:

$$\beta\text{-D-glucose} + O_2 \xrightarrow{\text{GOD}} \text{gluconolactone} + H_2O_2$$

$$\text{L-glutamate} + O_2 \xrightarrow{\text{GlOD}} \text{2-oxoglutarate} + NH_3 + H_2O_2, \text{ and}$$

$$\text{Acetylcholine} + O_2 + 2\,H_2O \xrightarrow{\text{AcE+ChO}} \text{betaine} + \text{acetate} + 2\,H_2O_2.$$

Each of the enzymes was loaded or coloaded onto a reactor (PTFE coil; 20 mm$\times$0.5 mm i.d.) packed with aminopropyl-CPG using glutaraldehyde as the cross-linking agent. The microdialysis system used is outlined in figure 1.2.

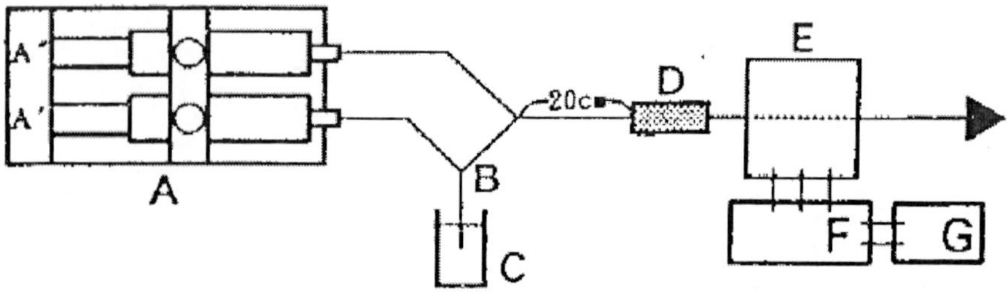

Figure 1.2. Schematic diagram of microdialysis system with immobilized enzyme reactor. A: two-channel microsyringe pump, B: microdialysis probe, C: sample, D: immobilized enzyme reactor, E: poly(1,2-diaminobenzene) film-coated platinum electrode, F: potentiostat, G: recorder. All the parts were connected with PTFE coil (i.d. 0.1 mm).

The apparatus consisted of a two-channel microsyringe pomp, potentiostat, and a strip chart recorder. Each of the immobilized enzyme reactors was positioned after a microdialysis probe, shown schematically in figure 1.3, and in front of a poly(1,2-diaminobenzene) film-coated platinum electrode. A constant potential (+0.6 V *vs.* Ag/AgCl) was applied to the polymer film-coated electrode and the current was recorded. The Ringer's solution was pumped at a constant flow (2 μ l min^{-1}) with a microsyringe pump.

For the glucose assay, a GOD immobilized reactor was used and 0.1 M phosphate buffer at pH 7.0 containing 1 mM sodium azide was selected as the carrier buffer. Under the recommended flow conditions, a linear calibration plot for the glucose assay was obtained over 2×10^{-6} – 1×10^{-2} M; the slope was 272 nA mM^{-1}.

For the L-glutamate assay, a GlOD immobilized reactor was used and 0.1 M phosphate buffer at pH 7.5 was selected as the optimum carrier buffer. Under the recommended flow conditions, the permeability of L-glutamate into the probe was 7.6 %. However, the conversion efficiency of L-glutamate to hydrogen peroxide was ca. 100 %. A linear calibration plot for L-glutamate assay was obtained over 2×10^{-6} – 5×10^{-3} M under the same flow conditions as the glucose assay; the slope was 181 nA mM^{-1}.

For the acetylcholine assay, a AcE+ChO coimmobilized reactor was used. However, choline produces hydrogen peroxide in this enzyme reactor and, therefore, interferes with the measurement of acetylcholine. Because of this, the ChO-catalase(CT) coimmobilized reactor was inserted in series before the AcE+ChO coimmobilized reactor to remove choline in the dialysate according to:

choline $\xrightarrow{\text{ChO+CT}}$ betaine + H_2O. When the borate buffer (0.1 M) at pH 8.5 was selected as the carrier buffer, the ChO-CT coimmobilized reactor had an activity that was sufficient to decompose choline below 2 mM. Also, the permeability of acetylcholine into the probe was about 11 %. A linear calibration plot for acetylcholine was obtained over $2\times10^{-6} - 5\times10^{-3}$ M; the slope was 82 nA mM^{-1}.

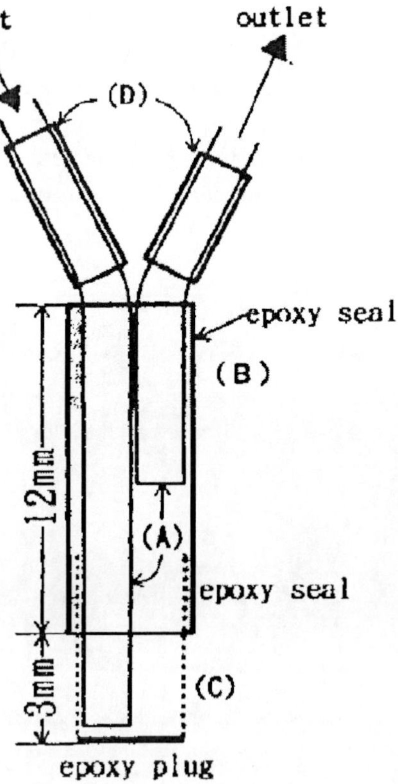

Figure 1.3. Schematic diagram of a microdialysis probe. A: fused silica tube (o.d. 150 μm, i.d. 75 μm), B: fused silica tube (o.d. 470 μm, i.d. 350 μm), C: dialysis fiber (o.d. 220 μm, i.d. 200 μm), D: stainless steel tube (o.d. 440 μm, i.d. 270 μm).

In this work, a poly(1,2-diaminobenzene) film-coated platinum electrode was used as an amperometric detector of the hydrogen peroxide generated enzymatically in the enzyme reactors. This polymer-coated platinum electrode prevents the electroactive interferents and proteins, commonly found in serum, from reaching the electrode surface and from interfering with the detection of the hydrogen peroxide generated in the enzyme reactor.

The *in vivo* monitoring of glucose in the rat brain was conducted by implanting a microdialysis probe into the frontal lobe (figure 1.4). The rat was allowed to recover overnight in a large plastic box with free access to food and water. The probe was then connected to a microsyringe pump through a swivel to allow the rat free movement. After 12 min, the signal current increased up to a steady value, corresponding to the glucose content in the rat brain. When a 1 ml aliquot of 20 %(w/v) glucose was injected into the abdomen of the rat, the signal current increased and then decreased to the value before glucose injection. It was found from these results that the present continuous flow method was useful for the *in vivo* monitoring of glucose. However, the sensitivity of the present method was not sufficient for an accurate *in vivo* monitoring of trace amounts of L-glutamate and acetylcholine in the rat brain.

Figure 1.4. *In vivo* assay.

1.3.1.2. An In Vivo Flow-Injection System for the Highly Selective and Sensitive Detection of L-Glutamate Using an Enzyme Reactor Involving Amplification [15-17]

L-glutamate is one of the most important excitatory neurotransmitters in the brain. Some enzyme-based biosensors have been proposed for the monitoring of the release of endogeneous glutamate into the extracellular space of the central nervous system. However, most of these methods were not sufficient in sensitivity for *in vivo* monitoring of L-glutamate while continuous calibrations of the sensor sensitivity during monitoring were not possible. Here, a calibration method is proposed to improve these problems by coupling microdialysate sampling in short intervals to a flow injection detection system with an enzyme reactor involving amplification by substrate recycling, which allows the alternate detection of standard and dialysate samples and enhancement of the sensitivity for L-glutamate. The system includes a microdialysis probe, two enzyme reactors, a six-way autoinjector, and a poly(1,2-diaminobenzene) film-coated platinum electrode (see figure 1.5). In the first stage (load for 3 min, figure 1.5) of the operation, the dialysate from the probe was delivered to the sample loop of the six-way autoinjector by perfusing the Ringer's solution at a flow rate of 2 μ l min^{-1}. Simultaneously, a 1 μ M L-glutamate standard solution (2.5 μ l) for checking drifts in the sensitivity was injected in the FIA line. The optimized carrier buffer was 0.1 M, pH 6.7, an ammonium phosphate buffer containing 0.5 mM NADH, 5 mM L-lysine, and 147 mM NaCl was pumped at 100 μ l min^{-1}. In the second stage (inject for 30 s, figure 1.5), the dialysate collected in the sample loop was injected into the FIA flow line and assayed electrochemically. The enzyme column E_2 was employed to enhance the sensitivity of L-glutamate as an on-line amplifier based on substrate recycling, as shown schematically in figure 1.6. To improve the substrate specificity of this reactor, acidic pretreatment and 6-diazo-5-oxo-L-norleucine treatment were carried out since trace amounts of NADH oxidase and glutaminase were contained as impurities in the commercially available GlOD. The resulting enzyme reactor worked as a specific reactor of L-glutamate involving amplification (with an amplification factor of about 60). The cycle was also initiated with 2-oxoglutarate, and thus enzyme column E_1 was positioned in series before enzyme column E_2 to eliminate enzymatically 2-oxoglutarate in the presence of excess of NADH and L-lysine in the carrier buffer, corresponding to:

$$\text{2-oxoglutarate} + \text{L-lysine} + \text{NADH} \rightarrow \text{saccharopine} + \text{NAD}^+.$$

This column E_1 can decompose completely and selectively 2-oxoglutarate of less than 0.3 mM added to the L-glutamate solutions when used as the eliminating column. By the present method, variations of the L-glutamate level released from the rat brain cells were assayed *in vivo*. The average amounts of L-glutamate in the rat brain was 6.2 μ M and the reproducibility of the measurement was 1.2 % r.s.d (n=7) under an analytical speed of 20 dialysates h^{-1}. This method was also applied to monitor the changes in the L-glutamate concentrations in the rat brain as a result of KCl stimulation, as shown in figure 1.7. As a result, the increase of the L-glutamate concentration in the rat brain was caused by the physiological action of KCl permeated from the dialysis probe into the extracellular space.

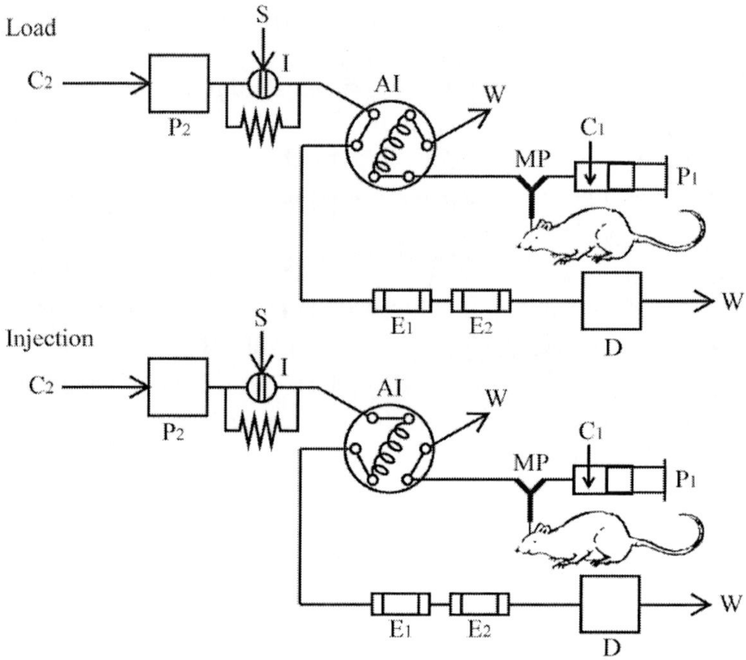

Figure 1.5. Two stages (load and inject) of an FIA system used for in vivo experimental of L-glutamate. P_1: syringe pump, P_2: double plunger μl pump, MP: microdialysis probe, AI: six-way autoinjector with a sample loop of 2.5 μl, I: injector with a sample loop of 2.5 μl, D: electrochemical detector with poly(1,2-diaminobenzene) film-coated platinum electrode, E_1: saccharopine dehydrogenase immobilized reactor, E_2: L-glutamate oxidase / glutamate dehydrogenase coimmobilized reactor, C_1: Ringer's solution, C_2: carrier buffer, S: standard L-glutamate solution(1 μM) to calibrate the detection system, W: waste.

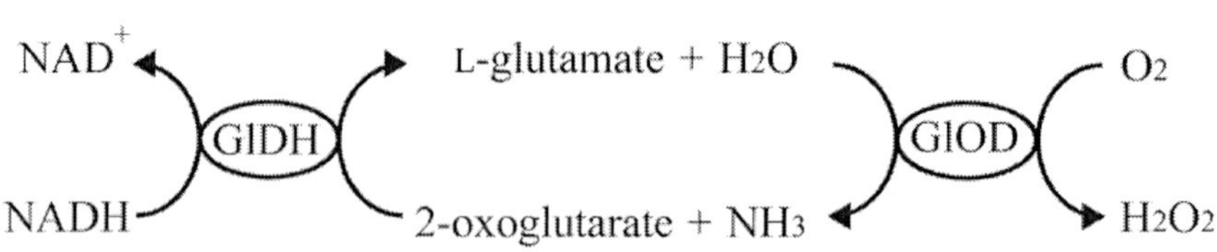

Figure 1.6. Enzymatic substrate recycling of L-glutamate in the L-glutamate oxidase (GlOD)/glutamate dehydrogenase (GlDH) coimmobilized reactor, under the conditions in the presence of excess of NADH, oxygen, and ammonium.

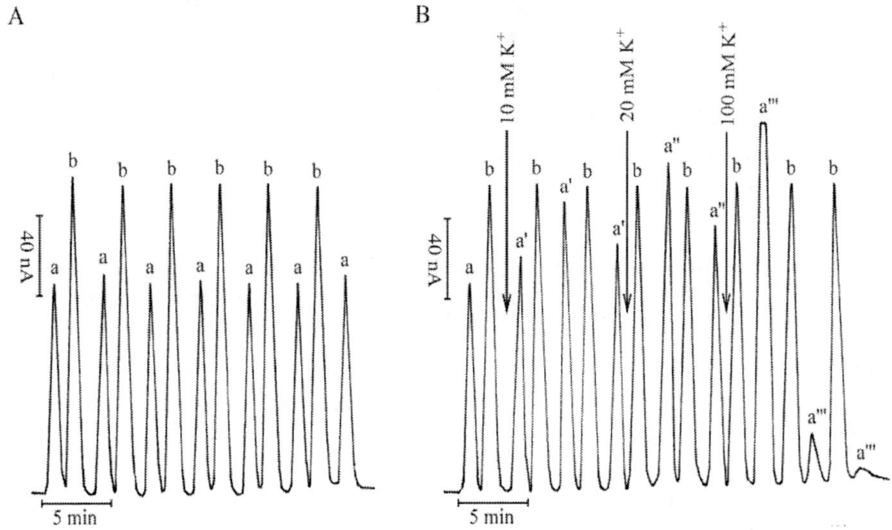

Figure 1.7. FIA signals of L-glutamate in the rat brain (A) and those of L-glutamate released from rat brain cells by KCl stimulation (B). KCl concentration in Ringer's solution a: 4 mM; a': 10 mM; a": 20 mM; a"': 100 mM; b: 1μM L-glutamate standard.

1.3.2. The Use of Biosensors as a Molecular Recognition Sensor

1.3.2.1. Simultaneous In Vivo Monitoring of Glucose, L-Lactate, and Pyruvate Concentrations in the Rat Brain [18, 19]

Most of the FIA methods with microdialysis sampling have generally been used for the determination of a single analyte. In many cases, however, it is preferable to have the opportunity to measure plural analytes from the dialysate sample, e.g., in the study of metabolic disorders. Glucose is a main nutrient in the brain, while L-lactate and pyruvate are found from glucose in an actively glycolyzing system under anaerobic and aerobic conditions, respectively, and are of interest in metabolic disorders. Therefore, the simultaneous on-line monitoring of glucose, L-lactate, and pyruvate in the brain would be of great benefit for studies on glucose metabolism in the brain.

The analytical system is similar to that shown in figure 1.5 and is based on the combination of a microdialysis sampling system and a flow-injection system with a triple enzyme electrode arranged perpendicularly for the flow direction. The triple enzyme electrode is constructed by hybridizing a poly(1,2-diaminobenzene) film to three sensing parts modified with the GOD, lactate oxidase (LOD), or pyruvate oxidase (POD) membrane cross-linked with gelatine using glutaraldehyde, which respond selectively to glucose, L-lactate, and pyruvate, respectively, without any cross-reactivity. The optimum carrier buffer 0.1 M phosphate buffer at pH 6.5 was selected and pumped at a flow rate of 0.2 ml min^{-1}. Also, when the flow rate of the Ringer's solution was 2 μl min^{-1}, the permeability into the probe was 13.3 % for glucose, 21.4 % for L-lactate, and 22.1 % for pyruvate, respectively. The differences of the permeability for three analytes are dependent on those in their molecular size, molecular structure, and hydrophilicity. Under these flow conditions, the optimum time intervals of the load and injection was 120 and 10 s, respectively, and up to 28 dialysates h^{-1} could be analyzed. The three sensing parts of the triple enzyme electrode responded linearly to the concentrations of glucose, L-lactate, and pyruvate between 0.1 and 10 mM, without any interference from oxidizable species (urate, ascorbate, and cysteine) present in serum and in the rat brain.

The proposed FIA method can be successfully applied to the simultaneous *in vivo* assay of three analytes in the rat brain, by implanting a microdialysis probe into the frontal lobe of the rat. The time courses of glucose, L-lactate, and pyruvate concentrations in the rat brain are shown in figure 1.8. The average amounts of glucose, L-lactate, and pyruvate in the extracellular space of the rat brain were about 0.8 mM each for glucose and L-lactate and less than 0.1 mM for pyruvate, respectively, although their concentrations varied much less with the elapse of time. When a 1 ml aliquot of 0.3 g ml^{-1} glucose was injected into the abdomen of the rat, the signal current of glucose increased up to a current corresponding to 1.7 mM, and then after 30 min it decreased, as shown in figure 1.8. Furthermore, a 1 ml aliquot of 0.6 g ml^{-1} glucose was injected similarly into the abdomen of the same rat, however, the glucose concentration in the rat brain was not increased beyond 1.8 mM. This means the presence of a "blood brain barrier" system, which is one of the physiological protection systems of the brain. However, the L-lactate and pyruvate concentrations, although the L-lactate concentration increased much less when the rat was in motion, were not influenced by glucose injections and remained almost constant during all periods of measurement.

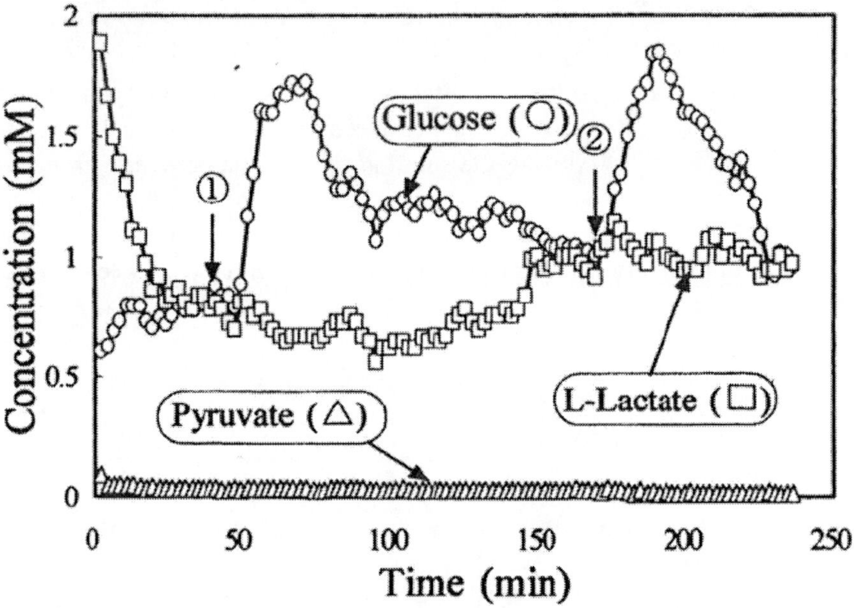

Figure 1.8. Time-course of glucose, L-glutamate, and pyruvate concentrations in rat brain. A 1.0 ml aliquot of 0.3 g ml^{-1} and 0.6 g ml^{-1} glucose solutions were injected into the abdomen of the same rat at Arrow ① and Arrow ②, respectively.

1.3.2.2. Microdialysis Fiber Enzyme Electrode for In Vivo Monitoring [20, 21]

Micro-flow injection biosensor systems with a microdialysate sampling were well suited for the on-line assay of a particular analyte in the dialysate with complex matrices. However, such a flow-system is not strictly a real-time monitoring process because of the delay time between sampling and detection. In this research, we prepared a microdialysis fiber enzyme electrode which makes possible real-time and *in vivo* monitoring. The electrode consists of a platinum fiber as a working electrode and a silver/silver chloride fiber as a reference electrode inserted into the tip and in the upper compartment of the dialysis fiber, respectively, as shown in figure 1.9. The platinum fiber was coated electrochemically with a poly(1,2-diaminobenzene) film, which permitted the selective detection of low molecular weight hydrogen peroxide generated enzymatically. We selected glucose electrodes using GOD as a typical model of the sensor of this type. The measurement was first done by holding stationary a soluble enzyme solution in the probe cavity, and then by the electrode with the enzyme layer immobilized on the polymer-coated platinum fiber. For the fiber electrode with the soluble enzyme, the response current was proportional to the concentration of glucose over the range of 5×10^{-6} – 1×10^{-3} M. This type of sensor has the advantages that the lifetime of the sensor can be extended by infusing in fresh enzyme solution when required and that a variety of sensors can be easily constructed by infusing the solution of the suitable hydrogen peroxide-producing enzymes without tedious immobilization procedures. In contrast, the sensitivity of the fiber electrode with the immobilized enzyme to glucose was about 83 % compared to that of the glucose electrode with the soluble enzyme.

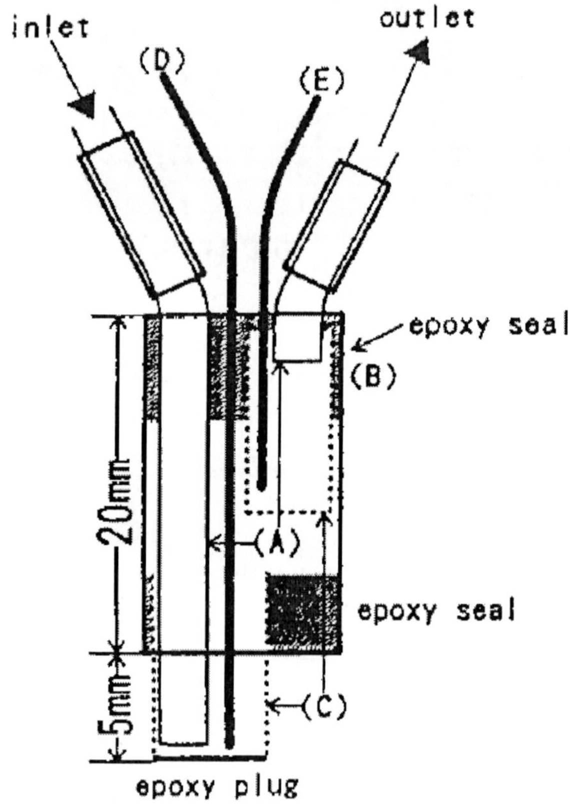

Figure 1.9. Schematic diagram of microdialysis electrode. A: fused silica tube (o.d. 105 µm, i.d. 40 µm), B: fused silica tube (o.d. 440 µm, i.d. 350 µm), C: dialysis fiber (o.d. 220 µm, i.d. 200 µm), D: platinum fiber (50 µm), E: silver/silver chloride fiber (50 µm).

1.4. General Aspects of Analytical Systems Based on Microfluidic Devices

1.4.1. Characteristics Provided by a Liquid Microspace

In general, microchannels fabricated on glass or other substrates can be considered as a "small reaction vessel"of a certain chemical reaction. In this case, the following three characteristics are very important in designing the device:

1. Short molecular diffusion distance
2. Large specific interfacial area
3. Small heat capacity

The above mentioned characteristic 1) means that the rapid mixing based on molecular diffusion is possible without mechanical mixing. The molecular diffusion time is given by:

$$t=L^2/D \tag{1}$$

where t, L, and D are the molecular diffusion time, diffusion distance and diffusion coefficient, respectively. As expressed in Eq. (1), the diffusion time is proportional to the square of the scale. Therefore, the diffusion time takes several hours to a day when the diffusion distance is 1 cm since the diffusion coefficient of typical molecular ions is on the order of 10^{-5} cm^2 / s. In contrast, it takes only several tens of seconds when the diffusion distance is 100 μm. This characteristic is very important when the certain analysis involving the slow diffusion of large molecules, e.g., immunoassay or other enzymatic methods.

Characteristic 2) is very advantageous for carrying out a chemical reaction or molecular transfer involving a solid-liquid or liquid-liquid interface. Concerning the solid-liquid interface, surface modification of the channel or packing of the chemically-modified beads allows efficient chemical reactions. On the other hand, a liquid-liquid interface formed inside a microchannel such as a multiphase laminar flow or liquid droplets can enhance the efficiency of the extraction without any mechanical mixing. The specific interfacial area, σ, can be expressed as:

$$\sigma = S/V \propto 1/L \tag{2}$$

where S and V are the interfacial area and the volume, respectively. Figure 1.10 shows the scale dependence of the molecular transportation time and the specific interface area. The specific interface area of the 100-μm scale microspace is equivalent to that provided by using a separatory funnel with rather vigorous mechanical shaking. These kinds of scale merits become remarkable below the scale of about 250 μm.

Characteristic 3) is also related to Characteristic 2) in the large specific interface inside a microchannel. Since the size of the space is rather small, a very fast response in the heating and cooling of the outer stimuli is expected. This characteristic is advantageous in rapid heat dissipation or temperature rise of the solution inside the microchannel, thus, an on-chip PCR system is actively undertaken by many researchers [22].

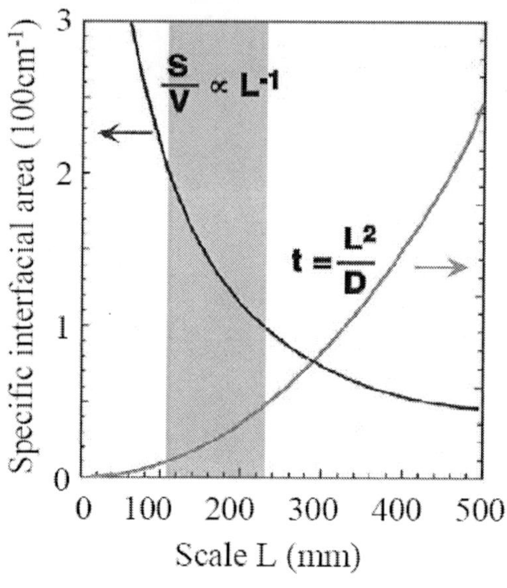

Figure 1.10. Scale dependence of specific interface area and diffusion time in microchip.

1.4.2. Multilayer Flow inside Microchannels

A low Reynolds number is one of the characteristics of the flow inside microchannels, resulting in the tendency for flow to be a laminar flow rather than turbulent flow [23-28]. Therefore, when two or more flow streams make contact at the confluence point, a relatively stable multilayer flow can be formed. Figure

1.11 shows a photograph of a typical organic-aqueous two layer flow formed inside a microchannel. Since the liquid-liquid interface is formed along the longitudinal direction, chasing the liquid-liquid extraction procedure can be carried out by scanning the detection point from upstream to downstream. Furthermore, by exploiting the liquid-liquid interface reaction, construction of the membrane structure inside the microchannel along the longitudinal direction is also possible. Applications using these techniques will be described in the next section.

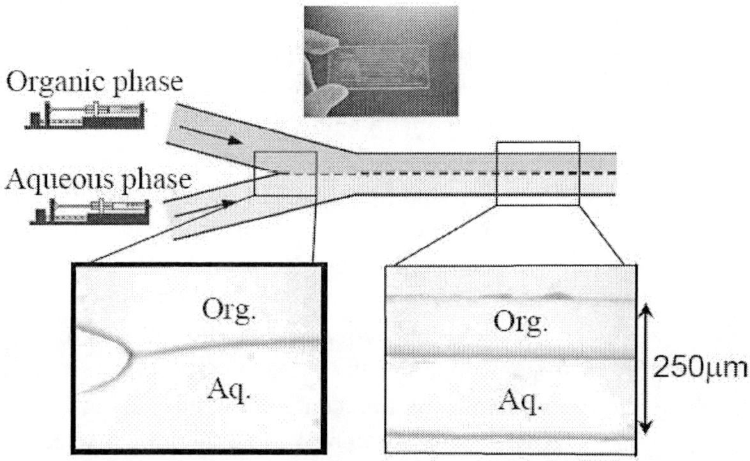

Figure 1.11. Typical photographs showing the liquid-liquid (organic-aqueous) interface formed inside a glass microchannel having 250μm width and 100μm depth.

1.5. Applications of Microfluidic Devices for Analytical Systems

1.5.1. Multi-Ion Analysis System Based on an Intermittent Reagent Pumping and Two Layer Flow Inside Microchannel [29, 30]

The first application of the microfluidic analytical device is a multi-ion optical sensing device. Conventional ion sensor technology is based on ion-ionophore complexation chemistry and membrane technology involving immobilization of the ion-sensing components such as an ionophore or a dye molecule. Basically, a single device can be applied to only one corresponding component. In addition, durability and low cost are required for the membranes, and more sophisticated ion-sensing technology is desired. Instead of a polymer membrane, we have used an organic/aqueous laminar two-phase flow as a medium for neutral ionophore-based ion pair extraction. By applying intermittent pumping of the multiple organic phases, which contain various ionophores, we have attempted to integrate multi-ion sensing onto a single microchip.

Figure 1.12 shows the basic concept of multi-ion sensing using a microchip. Different organic phases containing the same lipophilic pH indicator dyes but different ionophores are introduced sequentially into the microchannel by the on-off switching of syringe pumps. In this case, an organic phase without ionophores is introduced in between the two organic phase segments containing different ionophores, in order to avoid contamination. An aqueous sample solution containing different ions is introduced from the other inlet to form laminar two-phase flow with the segmented organic phases. The selective ion pair extraction reaction proceeds during the flow; thus, different ions can be selectively extracted into different organic phase segments depending on the selectivity of the neutral ionophores contained in the respective organic phases. Downstream in the flow, the ion pair extraction reaction becomes equilibrated, thus, downstream detection of color changes in the organic phase allows sequential and selective multi-ion sensing in a single aqueous sample solution containing multiple ions.

Valinomycin and DD16C5, which are known to exhibit high selectivity when used in conventional ion sensors [31–36], were selected as highly selective potassium and sodium ionophores, respectively.

Before investigating the intermittent reagent pumping technique, we investigated the position dependence of the signal intensity at the organic phase from the confluence point to the channel outlet. For the detection of the color change in the organic phase, we used a thermal lens microscope (TLM) which is known as a highly-sensitive detection technique [37,38]. The ion pair extraction reaction starts by contact between the organic and aqueous phases at the confluence point. Under continuous flow conditions, the TLM signal intensity increases, depending on the distance from the confluence point, as a result of the increase in the contact time. Therefore, position dependence of the TLM signal is directly correlated with the reaction profile inside the microchannel. Figure 1.13 shows the position dependence of the TLM signal intensity. As expected, increasing the contact distance ensured that the response was sensitive. At 100 mm from the confluence point, the response was saturated, indicating achievement of the ion pair extraction equilibrium. In this case, the corresponding contact time was 26.9 s. A roughly estimated diffusion coefficient by Fick's law of diffusion was ca. 10^{-5}-10^{-6} (cm^2/s), indicating that the response was mostly governed by molecular diffusion of the ion pair species in the organic phase. On the basis of these results, we concluded that a detection point more than 100 mm from the confluence point gives an equilibrium response under our experimental conditions. The response time obtained was in good agreement with the order-estimated value based on the molecular diffusion theory.

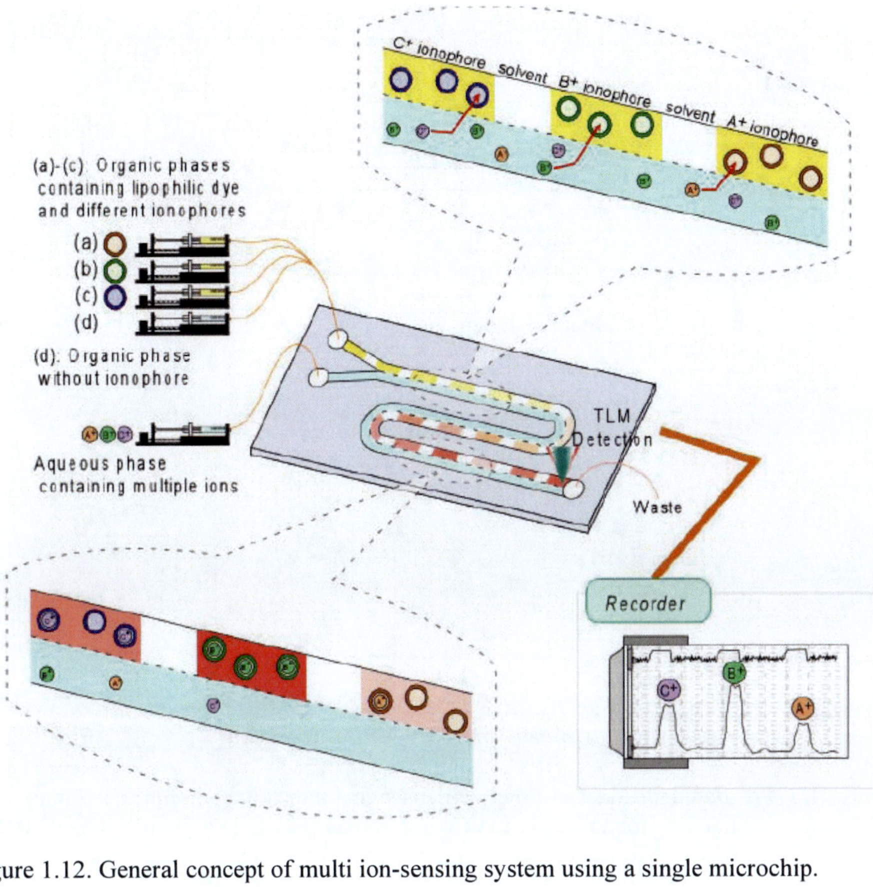

Figure 1.12. General concept of multi ion-sensing system using a single microchip.

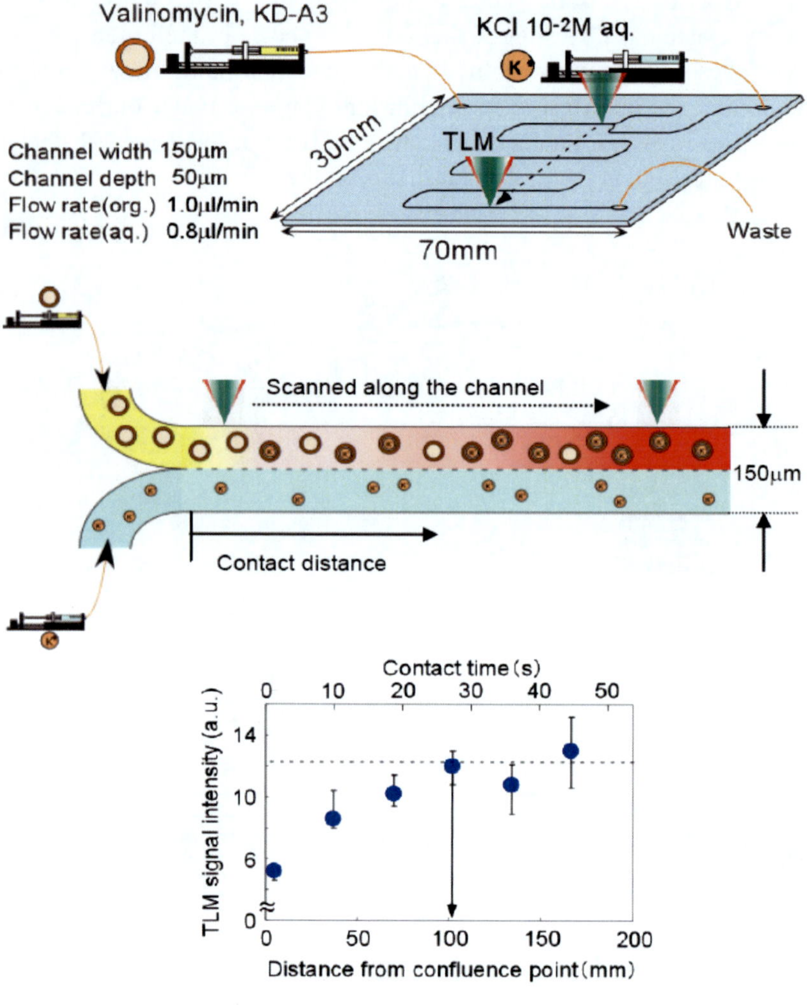

Figure 1.13. Position dependence of thermal lens signal intensity upon introduction of butyl acetate containing lipophilic pH indicator dye (KD-A3) and valinomycin, and 10^{-2}M potassium ion solution. Error bars indicate the variation in the TLM signal during measurement. Plots are the averaged signals obtained in the experiment.

Finally, we investigated three types of aqueous sample solutions, as shown in figure 1.14, to clearly show the demonstration of multiple-ion sensing using intermittent reagent pumping and a two-phase flow. When the aqueous phases containing a single type of ion were used, selective extractions occurred in each case; i.e., potassium ions were extracted only for the organic phase segment containing valinomycin, and sodium ions were extracted only for that containing DD16C5. When the aqueous phase containing both ions was examined, both ions were independently extracted into different organic phases, depending on the nature of the ionophores in the respective organic phases. The results successfully demonstrated our initial concept of realizing multi-ion determination by exploiting intermittent reagent pumping and the formation of a two-layer flow. The minimum volume of the single organic phase needed to obtain an equilibrium response without dilution by cross dispersion of the two organic phases was ca. 500 nL in the system, indicating that the required amounts of expensive reagents in one measurement could be reduced to a few nanograms.

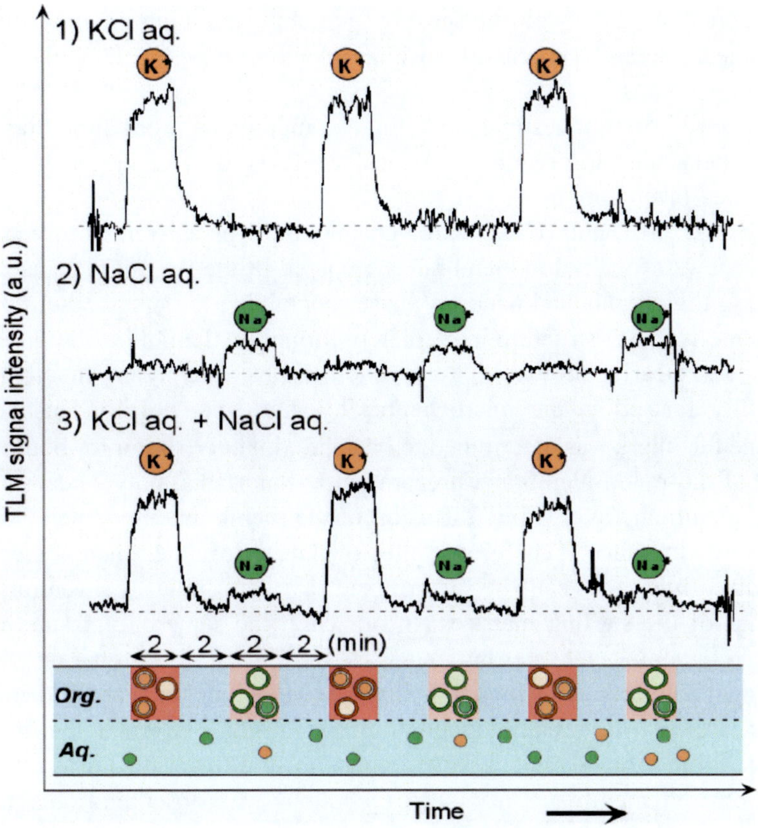

Figure 1.14. Response profiles for different aqueous solutions obtained by intermittent pumping of organic phases: 1) aqueous solution containing 10^{-2}M KCl, 2) aqueous solution containing 10^{-2}M NaCl, and 3) aqueous solution containing both 10^{-2}M KCl and 10^{-2}M NaCl. Introduction times of the organic phases were fixed at 2 min.

1.5.2. Biosensing System Based on an In Situ-Fabricated Organic Membrane Immobilizing Enzyme [39]

The second application is a novel biosensing system based on an in situ-fabricated membrane immobilizing enzyme. We have focused on the organic / aqueous interface formed inside the microchannel to realize the various types of analytical or synthetic applications. On the other hand, interfacial polymerization is a well known method for synthesizing a polyamide membrane at the organic / aqueous interface. Thus a combination of these technologies is expected to enable membrane formation inside the microchannel (see figure 1.15). Furthermore, surface modification of the polyamide membrane by the enzyme allows us development of a novel biosensing system involving complicated and sophisticated chemical processes such as membrane permeation and selective enzymatic reactions.

Here, we first investigated a nylon membrane formation inside the microchannel under an organic / aqueous two-layer flow or organic / aqueous / organic three-layer flow.

Figures 1.16 (a) and (b) show the channel patterns as well as top and cross-sectional views of the nylon membranes prepared inside the microchannels. After introducing the organic and aqueous phases containing the membrane monomers, a membrane-type 3D-structure immediately formed at the liquid / liquid interface position. As clearly shown in these photographs, the nylon membrane was successfully formed in the microchannel by interfacial polycondensation. The membrane thickness was determined to be approximately 10 μm for both types. In the case of 6,6-nylon membrane preparation by interfacial polycondensation, the prepared membrane avoids interdiffusion of the membrane monomers across the liquid/liquid interface. Therefore, the thickness of the membrane at the liquid/liquid interface can be determined by the initial concentration and reactivities of the starting matrices. In our case, the same organic and aqueous solutions containing the starting matrices were used for preparing single or parallel dual membranes. Although linear flow velocities for these two conditions were not exactly the same, monomer concentration governed the membrane thickness. From cross-sectional photographs, no significant thickness distribution of the prepared membrane from top to bottom was observed. This fact also supports the explanations given above.

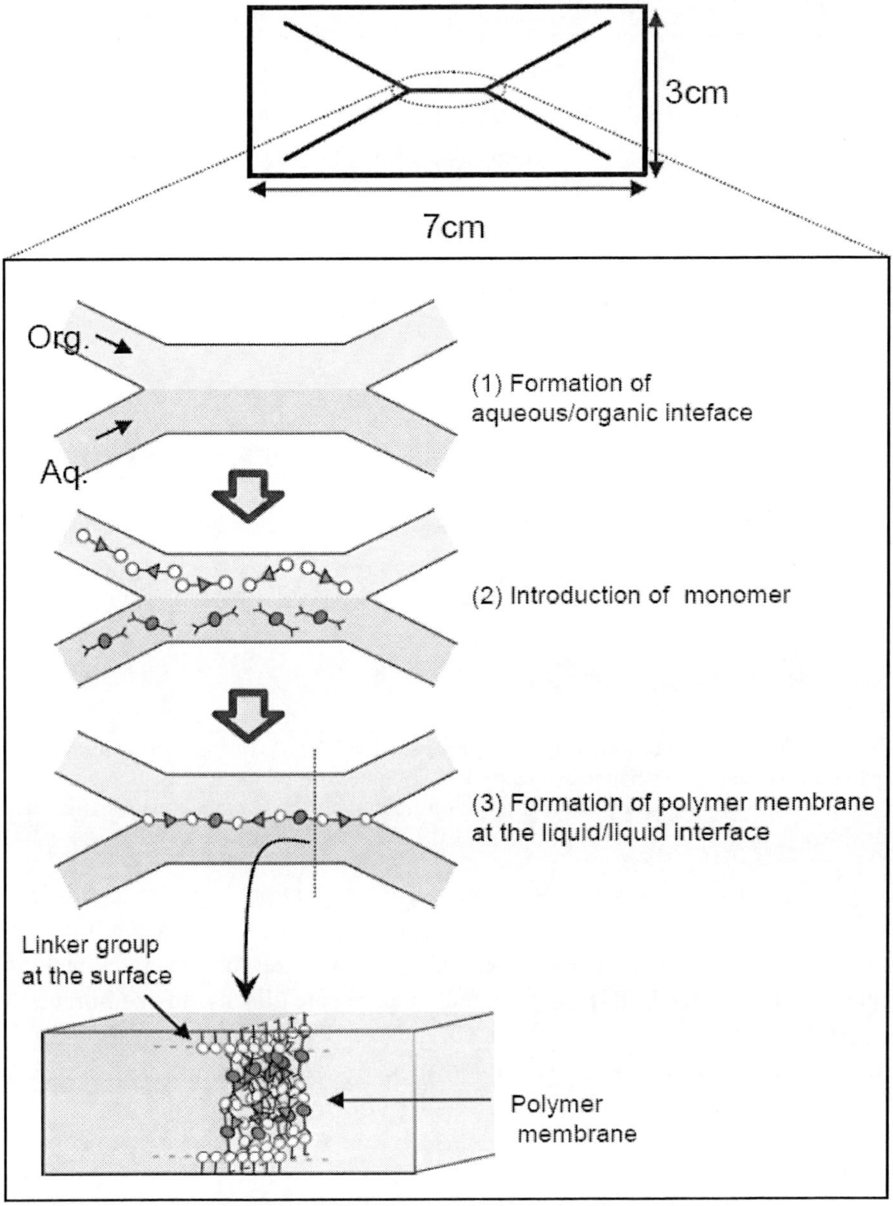

Figure 1.15. General concept of polymer membrane formation under an organic/aqueous two-phase flow in an X-shaped microchannel.

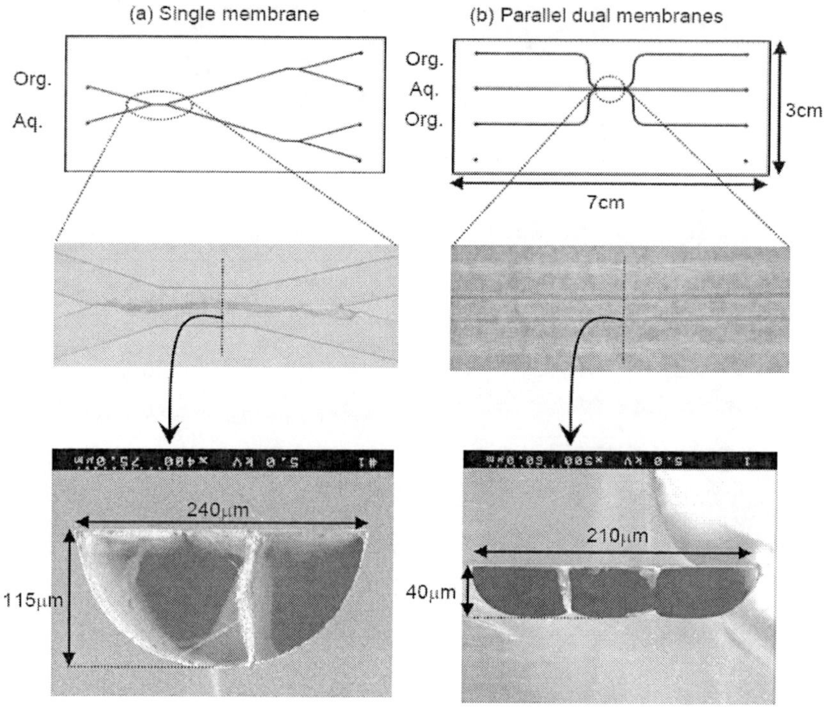

Figure 1.16. Channel patterns and top and cross-sectional views of the nylon membrane prepared inside the microchannel. (a) Single membrane formed under organic/aqueous two-layer flow; (b) Parallel dual membranes formed under organic/aqueous/organic three-layer flow.

In order to confirm the molecular transport function of the membrane, we carried out a simple permeation experiment (figure 1.17(a)) using the nylon membrane prepared inside the microchannel. In this case, the generated ammonia species (NH_3 gas, NH_4OH) is expected to permeate the nylon membrane and change the pH of the phenolphthalein phase. Therefore, permeation of the ammonia species through the 6,6-nylon membrane can be detected as a color change of the phenolphthalein phase by using TLM.

Figure 1.17(b) shows the time course of the TLM signal generated at the phenolphthalein phase when ammonium chloride and sodium hydroxide solutions were introduced into the other side of the channel separated by the membrane. In order to confirm leaking of the sodium hydroxide solution to the phenolphthalein phase through the membrane, the introduction of sodium hydroxide solution without ammonium chloride was also investigated. In both cases, TLM measurements were performed at the phenolphthalein phase under the stopped flow condition.

As can be seen, the TLM signal at the phenolphthalein phase did not change when sodium hydroxide solution was introduced without ammonium chloride, while on the other hand, the signal rapidly increased when the ammonia species was generated at the other side of the membrane. These results clearly suggest that the nylon membrane prepared inside the microchannel successfully worked as a permeation membrane of the ammonia species.

In figure 1.17 (b), the TLM signal reached a steady-state at ca. 1 minute. In our experiment, the response time is mainly governed by diffusion of the ionic species (NH_4^+, OH^-) in the liquid phase rather than that of the ammonia species in the nylon membrane due to the large difference in diffusion coefficients. It is well known that the diffusion time of small chemical species in liquid is much slower (generally $\sim 10^{-5}$ cm^2/sec in diffusion coefficient) compared to that of gas species (generally $\sim 10^{-1}$ cm^2/sec in diffusion coefficient). Based on the Fick's law of diffusion, the diffusion time of small molecules for a hundred micrometer-sized channel is approximately several tens of seconds. In our case, we prepared a highly permeable nylon membrane and the color change (deprotonation) of phenolphthalein is considered as a diffusion-controlled reaction, so that our experimental results corresponded well with the diffusion theory. For membrane permeation-based chemical processes, a micrometer-sized channel is quite advantageous for obtaining rapid equilibrium.

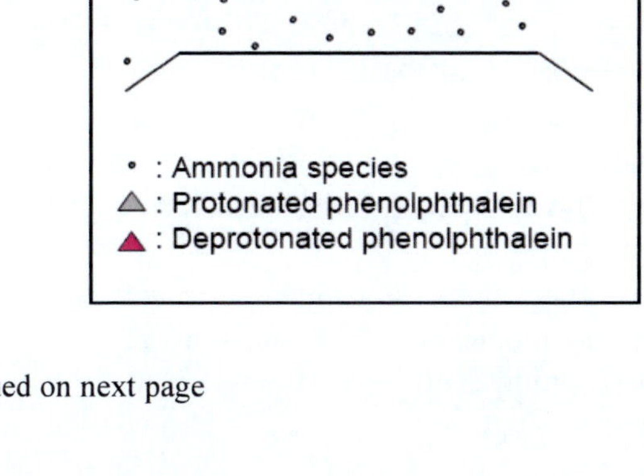

Figure 1.17. Continued on next page

(b)

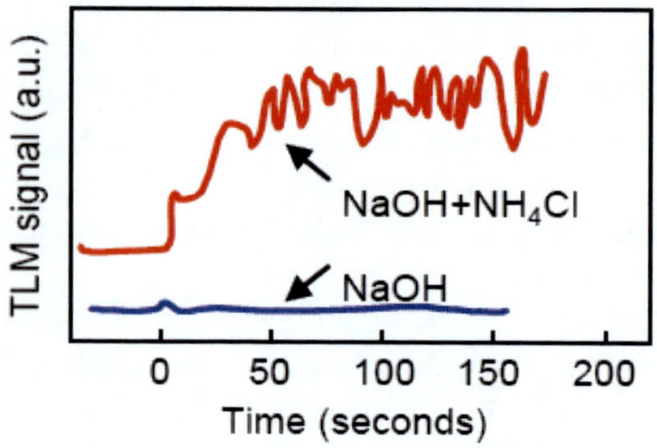

Figure 1.17. Ammonia species permeation experiment. (a) Schematic illustration of experimental system; (b) Time course of the TLM signal just after stopping the syringe pump.

Finally, we applied this membrane system to a novel biosensing system by surface modification with an enzyme. We chose horseradish peroxidase as an immobilizing enzyme and hydrogen peroxide as a nylon membrane-permeable substrate, respectively. Since the nylon membrane has residual amine groups, a well-known immobilization procedure using glutaraldehyde and sodium borohydride can be applied. For biosensing, horseradish peroxidase catalyses organic dye formation by using hydrogen peroxide, N-ethyl-N-(2-hydroxy-3-sulfopropyl)-m-toluidine (TOOS), and 4-aminoantipyrine (4-AAP) as substrates. Figure 1.18 (a) illustrates the experiment. Horseradish peroxidase was immobilized on one side of the membrane surface and hydrogen peroxide was supplied from the other side. The enzyme-immobilized side of the channel was filled with TOOS and 4-AAP. When the flow stopped, the TLM signal increased rapidly and reached a steady-state at approximately 250 seconds ((A) in figure 1.18 (b)). In this case, no signal was obtained by introduction of TOOS and 4-AAP without hydrogen peroxide (Control 1 in figure 1.18 (b)). Since the TLM signal obtained when mixing all the substrates inside the bare channel (without membrane-immobilized enzyme) was rather slow (Control 2 in figure 1.18 (b)), a rapid increase of the TLM signal shown in figure 1.18 (b) shows that hydrogen peroxide permeation and subsequent enzyme reaction at the membrane surface were successfully completed.

Since one of the substrates, TOOS, is ionic for its sulfonate group, the produced dye molecule should also have an ionic group. As mentioned in the ammonia species permeation experiment, the nylon membrane prepared here did not easily transport hydrophilic species under normal pressure conditions. When the detection point of TLM was shifted to the other side separated by the membrane ((B) in figure 1.18 (b)), no signal was observed, supporting the finding that the colored product did not permeate through the membrane.

These results suggest that the development of a novel biosensing system involving complicated and sophisticated chemical processes such as membrane permeation and selective enzymatic reaction was successfully achieved by using an in situ-fabricated membrane and surface modification chemistry.

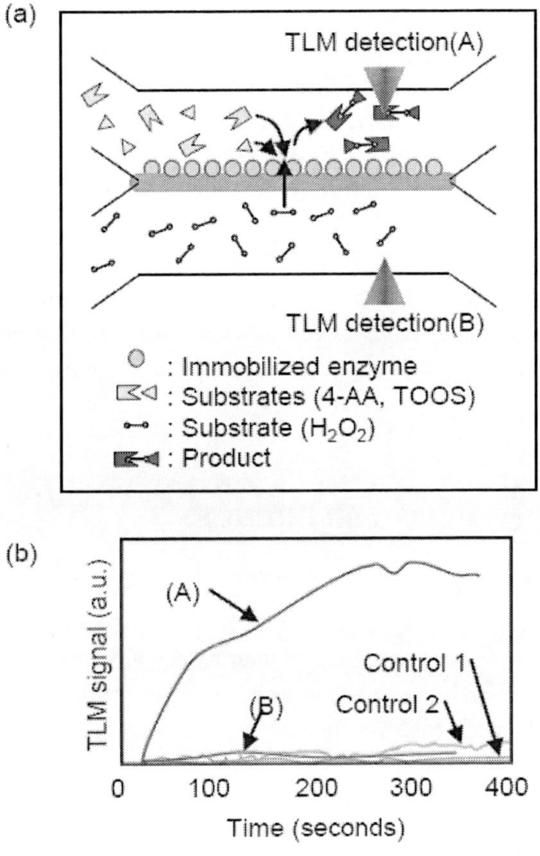

Figure 1.18. Substrate permeation and subsequent enzyme reaction experiment. (a) Schematic illustration of experiment system. (b) Time course of the TLM signal just after stopping the syringe pump. TLM detection points A and B indicate detection points in the microchannel separated by the innerchannel membrane, shown in (a). (A) Enzyme immobilized side. (B) Hydrogen peroxide supplying side. Controls 1 and 2 were obtained at detection point A.

1.6. Conclusion

The proposed micro-flow injection biosensor systems with an on-line microdialysis sampling are very selective and useful for the simultaneous *in vivo* monitoring of biomolecules without any interference from oxidizable species and proteins present in the extracellular space of the rat brain. Furthermore, these methods can be expected to be applied to the *in vivo* monitoring of a variety of biomolecules in other living cells such as the blood stream, subcutaneous tissue, and muscle.

Concerning the microfluidic chip systems, both devices introduced here effectively employed the specific characteristics provided by liquid microspace and multilayer flow formation inside a microchannel. The first application to an ion analysis device allowed multi-ion sensing, which was not easily demonstrated by conventional ion sensor technology using a solvent polymeric membrane. In addition, durability and the high cost of membrane-based sensors could be avoided. The second application of in situ-fabricated membrane-based biosensing may allow complicated and sophisticated chemical processes such as membrane permeation and enzymatic reaction to be realized. Designing novel microfluidic devices employing characteristics of liquid microspace and microflow is expected to be instrumental in the development of practical and useful bioanalytical tools in the near future.

References

[1] Robinson, E.T; Justice Jr, J.B. (1991). Microdialysis in the Neuroscience. Amsterdam: *Elsevier.*

[2] Kaptein, W.A; Zwaagstra, J.J; Venema, K; Korf, J. (1998). Continuous ultraslow microdialysis and ultrafiltration for subcutaneous sampling as demonstrated by glucose and lactate measurements in rats. *Anal. Chem.,* 70, 4696-4700.

[3] Moscone,D; Pasini, M; Mascini, M. (1992). Subcutaneous microdialysis probe coupled with glucose biosensor for in vivo continuous monitoring. *Talanta, 39*, 1039-1044.

[4] Zilkha, E; Koshy, A; Obrenovitch, T.P; Bennetto, H.P; Symon, L. (1994). Amperometric biosensors for on-line monitoring of extracelluar glucose and glutamate in the brain. *Anal. Lett.,* 27, 453-473.

[5] Palmisano, F; Santis, A.De; Tantillo, G; Volpicella, T; Zambonin, P.G. (1997). Microbial detection by a glucose biosensor coupled to a microdialysis fiber. *Analyst,* 122, 1125-1128.

[6] Niwa, O; Torimitsu, K; Morita, M; Osborne, P; Yamamoto, K. (1996). Concentration of extracellular L-glutamate released from cultured nerve cells measured with a small-volume online sensor. *Anal. Chem.,* 68, 1865-1870.

[7] Niwa, O; Horiuchi, T; Torimitsu, K. (1997). Continuous monitoring of L-glutamate released from cultured nerve cells by an online sensor coupled with micro-capillary sampling. *Biosens. Bioelectron.,* 12, 311-319.

[8] Berners, M.O.M; Boutelle, M.G; Fillenz, M.(1994). On-line measurement of brain glutamate with an enzyme / polymer-coated tubular electrode. *Anal. Chem.,* 66, 2017-2021.

[9] Niwa, O; Horiuchi, T; Kurita, R; Torimitsu, K. (1998). On-line electrochemical sensor for selective continuous measurement of acetylcholine in cultured brain tissue. *Anal. Chem.,* 70, 1125-1132.

[10] Fang, Q; Shi, X.T; Sun, Y.Q; Fang, Z.L. (1997). A flow injection microdialysis sampling chemiluminescence system for in vivo on-line monitoring of glucose in intravenous and subcutaneous tissue fluid microdialysates. *Anal. Chem.,* 69, 3570-3577.

[11] Dittrich, P. S.; Tachikawa, K.; Manz, A. (2006). Micro Total Analysis Systems. Latest Advancements and Trends. *Anal. Chem.,* 78, 3887-3908.

[12] Vilkner, T.; Janasek, D.; Manz, A. (2004). Micro Total Analysis Systems. Recent Developments. *Anal. Chem.,* 76, 3373-3386.

[13] Manz, A.; Grabner, N.; Widmer, H. M. (1990). Miniaturized total chemical analysis systems: a novel concept for chemical sensing. *Sens. Actuators B,* 1, 244–248.

[14] Yao, T; Suzuki, S; Nishino, H. (1995). On-line amperometric assay of glucose, L-glutamate, and acetylcholine using microdialysis probes and immobilized enzyme reactors. *Electroanalysis,* 7, 1114-1117.

[15] Yao,T; Suzuki, S; Nishino, H. (1998). Highly sensitive detection of L-glutamate by on-line amperometric micro-flow analysis based on enzymatic substrate recycling. *Talanta,* 45, 917-923.

[16] Yao, T; Nanjyo, Y; Nishino, H. (2001). Micro-flow in vivo analysis of L-glutamate with an on-line enzyme amplifier based on substrate recycling. *Anal. Sci.,* 17, 703-708.

[17] Yao, T; Nanjyo, Y; Tanaka, T; Nishino, H. (2001). An electrochemical in vivo flow-injection system for highly selective and sensitive detection of L-

glutamate using enzyme reactor involving amplification. *Electroanalysis,* 13, 1361-1366.

[18] Yao, T; Yano, T. (2004). On-line microdialysis assay of L-lactate and pyruvate in vitro and in vivo by a flow-injection system with a dual enzyme electrode. *Talanta,* 63, 771-775.

[19] Yao, T; Yano, T; Nishino, H. (2004). Simultaneous in vivo monitoring of glucose, L-lactate, and pyruvate concentrations in rat brain by a flow-injection biosensor system with an on-line microdialysis sampling. *Anal. Chim. Acta,* 510, 53-59.

[20] Yao, T; Suzuki, S; Nishino, H. (1997). Microdialysis fiber electrode for glucose with soluble and immobilized glucose oxidase. *Anal. Sci.,* 13, 665-667.

[21] Yao, T; Suzuki, S; Nishino, H. (1997). Microdialysis fiber enzyme electrode involving amplification by substrate recycling. *Electroanalysis*, 9, 950-951.

[22] Roper, M. G.; Easley, C. J.; Landers, J. P. (2005). Advances in Polymerase Chain Reaction on Microfluidic Chips. *Anal. Chem.,* 77, 3887-3894.

[23] Hibara, A.; Tokeshi, M.; Uchiyama, K.; Hisamoto, H.; Kitamori, T. (2001). Integrated Multilayer Flow System on a Microchip. *Anal. Sci.,* 17, 89-93.

[24] Kim, H.-B.; Ueno, K.; Chiba, M.; Kogi, O.; Kitamura, N. (2000). Spatially-Resolved Fluorescence Spectroscopic Study on Liquid/Liquid Extraction Processes in Polymer Microchannels. *Anal. Sci.,* 16, 871-876.

[25] Kamholz, A. E.; Weigl, B. H.; Finleyson, B. A.; Yager, P. (1999). Quantitative Analysis of Molecular Interaction in a Microfluidic Channel: The T-Sensor. *Anal. Chem.,* 71, 5340–5347.

[26] Brody, J. P.; Yager, P. (1997). Diffusion-based extraction in a microfabricated device. *Sens. Actuators A,* 58, 13-18.

[27] Weigl, B. H.; Yager, P. (1999). Microfluidic Diffusion-Based Separation and Detection. *Science,* 283, 346–347.

[28] Kenis, P. J. A.; Ismagilov, R. F.; Whitesides, G. M. (1999). Microfabrication Inside Capillaries Using Multiphase Laminar Flow Patterning. *Science,* 285, 83-85.

[29] Hisamoto, H.; Horiuchi, T.; Tokeshi, M.; Hibara, A.; Kitamori, T. (2001). On-Chip Integration of Neutral Ionophore-Based Ion Pair Extraction Reaction. *Anal. Chem.,* 73, 1382–1386.

[30] Hisamoto, H.; Horiuchi, T.; Uchiyama, K.; Tokeshi, M.; Hibara, A.; Kitamori, T. (2001). On-Chip Integration of Sequential Ion-Sensing System Based on Intermittent Reagent Pumping and Formation of Two-Layer Flow. *Anal. Chem.,* 73, 5551-5556.

[31] Suzuki, K.; Sato, K.; Hisamoto, H.; Siswanta, D.; Hayashi, K.; Kasahara, N.; Watanabe, K.; Yamamoto, N.; Sasakura, H. (1996). Design and Synthesis of Sodium Ion-Selective Ionophores Based on 16-Crown-5 Derivatives for an Ion-Selective Electrode. *Anal. Chem.,* 68, 208-215.

[32] Hisamoto, H.; Satoh, S.; Satoh, K.; Tsubuku, M.,; Siswanta, D.; Shichi, Y.; Koike, Y.; Suzuki, K. (1999). Theory and Practice of Rapid Flow-Through Analysis (RFA) 2; RFA Based on Ion-Selective Optode Detection and its Application to Sodium Ion Determination. *Anal. Chim. Acta,* 396, 131–141.

[33] Hisamoto, H.; Kim, K.-H.; Manabe, Y.; Sasaki, K.; Minamitani, H.; Suzuki, K. (1997). Ion Sensitive and Selective Active Waveguide Optodes. *Anal. Chim. Acta,* 342, 31–39.

[34] Kurihara, K.; Ohtsu, M.; Yoshida, T.; Abe, T.; Hisamoto, H.; Suzuki, K. (1999). Micrometer-Sized Sodium Ion-Selective Optodes Based on a "Tailed" Neutral Ionophore. *Anal. Chem.,* 71, 3558–3566.

[35] Suzuki, K.; Ohzora, H.; Tohda, K.; Miyazaki, K.; Watanabe, K.; Inoue, H.; Shirai, T. (1990). Fibre-optic potassium ion sensors based on a neutral ionophore and a novel lipophilic anionic dye. *Anal. Chim. Acta,* 237, 155–164.

[36] Hisamoto, H.; Miyashita, N.; Watanabe, K.; Nakagawa, E.; Suzuki, K. (1995). Ion Sensing Film Optodes: disposable ion sensing probes for the determination of Na^+, K^+, Ca^{2+} and Cl^- concentrations in serum. *Sens. Actuators B,* 29, 378–385.

[37] Tokeshi, M.; Uchida, M.; Hibara, A.; Sawada, T.; Kitamori, T. (2001). Determination of Subyoctomole Amounts of Nonfluorescent Molecules Using a Thermal Lens Microscope: Subsingle-Molecule Determination. *Anal. Chem.,* 73, 2112-2116.

[38] Tokeshi, M.; Yamaguchi, J.; Hattori, A.; Kitamori, T. (2005). Thermal Lens Micro Optical Systems. *Anal. Chem.,* 77, 626-630.

[39] Hisamoto, H.; Shimizu, Y.; Uchiyama, K.; Tokeshi, M.; Kikutani, Y.; Hibara, A.; Kitamori, T. (2003). Chemico-Functional Membrane for Integrated Chemical Processes on a Microchip. *Anal. Chem.,* 75, 350-354.

In: Environmentally Harmonious Chemistry... ISBN: 978-1-60876-428-0
Editors: M. Anpo and K. Mizuno

Chapter 2

DEVELOPMENT AND APPLICATION OF FUNCTIONAL INORGANIC MATERIALS

Masahiro Tatsumisago
Department of Applied Materials Science, Graduate School of Engineering, Osaka Prefecture University, Sakai, Osaka 599-8531, Japan

1. INTRODUCTION

Textbooks on human history usually start from the "Stone Age," "Bronze Age," "Iron Age," and so on through the ages. As can be seen, it is interesting to note that human history is classified by the materials which were originally used at that time. Today, we may still be in the "Iron Age" or we could be in the "Carbon and Silicon Age." We can recognize the importance of material science from the fact that the practical innovation of materials can strongly influence human history. Our modern life is based on a variety of functional materials with electric, dielectric, optic, magnetic, thermal, mechanical, and chemical properties. These materials make our everyday life more convenient.

How can we successfully change raw materials to such functional materials? In order to do so, it is important to know why such properties appear in materials. Table 1 lists the various functions observed in materials, the origin of these functions and examples of functional materials and products. The origin of the functions should be considered every time when we design new functional materials from raw materials. Functional materials are usually obtained by some modification of raw materials based on the origin of the functions. Modifications include processes and treatments to change the state or shape of the raw materials. For example, optical fiber is shown as a product in optical materials in Table 1. An optical fiber is a transparent fiber designed to guide light along its length by confining as much light as possible in a propagating form. Optical fibers are widely used in fiber-optic communication, which permits transmission over long distances. Is the modification in this case only the change in the shape? In addition to the fibrous shape, extremely high transparency is required for optical fiber materials. Generally, ultra-pure silica glass with no absorption and scattering centers in a given wavelength range is used for optical fiber materials. The functional material of ultra-pure silica glass is obtained by some modification of a raw material, crude silica. In the case of optical fibers, the modifications are not only fiber drawing but purification and vitrification.

Table 1. Functions in materials

Functions	Properties	Origin of functions	Examples of functional materials and products
Electrical	Electrical conduction	Ion or electron transfer	Semiconductor, Solid electrolyte, Superconductor
Magnetic	Diamagnetism, Ferromagnetism	Arrangement of spin	Ferrite, Rare-earth magnet
Surface	Adsorption on solid surface	Surface energy of solid	Catalyst, Gas sensor, Fine particles
Optical	Absorption of light, Emission of light	Transparency, Energy conversion	Optical fiber, Laser
Dielectric	Polarization, Ferroelectrics	Local motion of ions	$BaTiO_3$, PZT
Thermal	Thermal expansion, Thermal conduction	Lattice vibration,	Thermal insulator, Thermal conductor
Mechanical	Elasticity, Malleability, Hardness	Chemical bond, Defect, Dislocation, Grain boundary	Structure materials

In this chapter, we examine how raw materials are changed or modified into functional materials. Alumina, Al_2O_3, will be taken up as an example of such raw materials. We will demonstrate how it is modified into a variety of functional materials used for jewelry, insulators, porous powders, super-repellent films, etc. This chapter also reports the development and application of superionic conducting glassy materials as one of the most promising functional materials for all-solid-state batteries.

2. FUNCTIONAL MATERIALS FROM ALUMINA

In the previous section, it was mentioned that various functional materials are obtained by some modifications such as single crystal growth, vitrification, fiber drawing, film formation, and other modifications of raw materials such as simple oxides. Figure 1 shows the various modifications of Al_2O_3 into a variety of functional materials and products. It is interesting to note that many important products which are indispensable to our everyday life are obtained from alumina by several modifications. Ruby is a single crystal of alumina and the beautiful red color comes from the doping of small amounts of Cr^{3+} ions. Ceramic substrates for integrated circuit film are polycrystalline alumina sheets and these products are prepared by the sintering of molded alumina powder. Alumina fibers are obtained by a spinning process of alumina precursor solutions; in this case they are used in fiber-reinforced composites. Alumina catalyst carriers are porous materials in which the specific surface area is extremely high and as such they can be impregnated with large amounts of catalysts. Various modifications like doping, shaping, single crystal growth, and other changes in the raw material of simple Al_2O_3 powders to various functional materials such as jewels, insulating sheets, fibers, and catalyst carriers are possible, as seen in figure 1. These alumina-based functional materials support our convenient modern life.

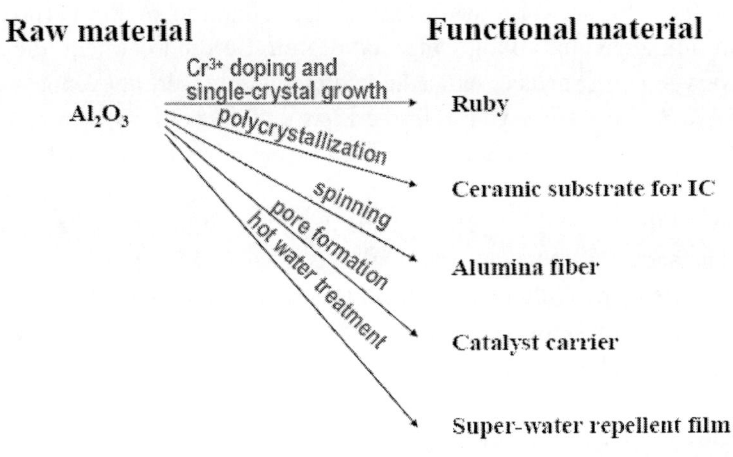

Figure 1. Modifications of raw material Al_2O_3 into a variety of functional materials and products.

The last example of the modification of alumina in figure 1 is hot water treatment. This treatment leaves ultra-water-repellent coatings on various substrates such as glasses, metals and plastics, and this technology has been recently developed to obtain nanocrystal-precipitated thin films. Common examples of water repellency include cookware such as Teflon-coated pots and pans, water resistant sprays for clothing and umbrellas, and solutions applied to windshields to repel water droplets. Many people are also familiar with the example in nature of water beading on the leaves of the lotus plant.

Wettability of a solid surface is governed not only by the chemical properties of the surface such as hydrophobic and hydrophilic effects but also by the microstructure of the surface. The lotus leaf's water repellency is mainly due to its surface structure. A surface with tiny bumps and unevenness in the microstructure has greater water repellency than a smooth surface. The water-repellent and hydrophilic properties are thus enhanced by the fine roughness. Modification with hot water treatment to alumina is to generate such fine roughness on its surface. Alumina thin films with a roughness of several ten nanometers are formed by immersing the porous alumina gel films prepared by the sol-gel method in hot water. Starting with the compound named aluminum tri-sec-butoxide dissolved in isopropyl alcohol, the alkoxide is hydrolyzed and polycondensed to produce a coating solution. Glass plates are dipped in this solution and then withdrawn, followed by 10 min of heat treatment at 400 °C. This produces a non-crystalline alumina gel film. Figure 2 shows a FE-SEM photograph of the surface of the alumina film immersed in boiling water for 10 min. Before treatment, the alumina film has a very smooth surface. After the treatment, however, unevenness of 20 to 50 nm petal structure is observed, referred to as a "flower-like structure"[1]. This is probably due to the formation of pseudoboehmite nanocrystals with the dissolution-reprecipitation process by the reaction between the sol-gel derived porous alumina films and H_2O. This flower-like microstructure is very similar to that of the surface of a lotus leaf. Fluoroalklylsilanes (FAS), which are known to be excellent water-repellent agents, are coated on the flower-like alumina film to show extremely high water repellency with a contact angle for water of about 165 degrees. A photograph of a water droplet of about 1 mm in diameter on the FAS-coated flower-like alumina film is shown in figure 3. The surface is so water-repellent that the droplets roll off very quickly. Because this film appears smooth when viewed with the naked eye, the water droplet's reflection can be seen in the photograph. The surface structure of the film, which is on a nanometer scale, is much smaller than the wavelength of visible light. Consequently, the structure does not scatter visible light, making the film transparent to the naked eye.

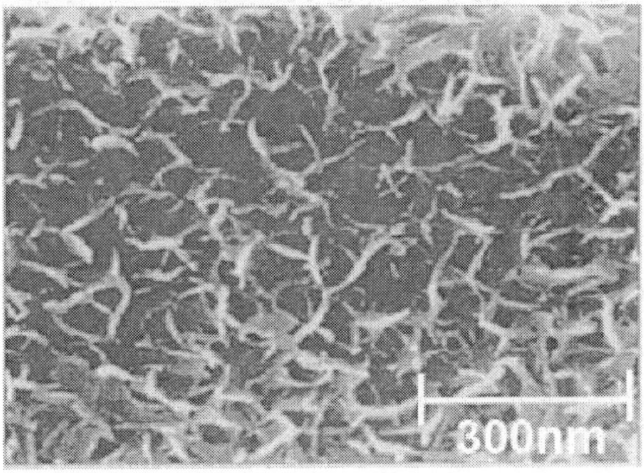

Figure 2. Field emission type scanning electron micrograph of the surface of the alumina film immersed in boiling water for 10 min.

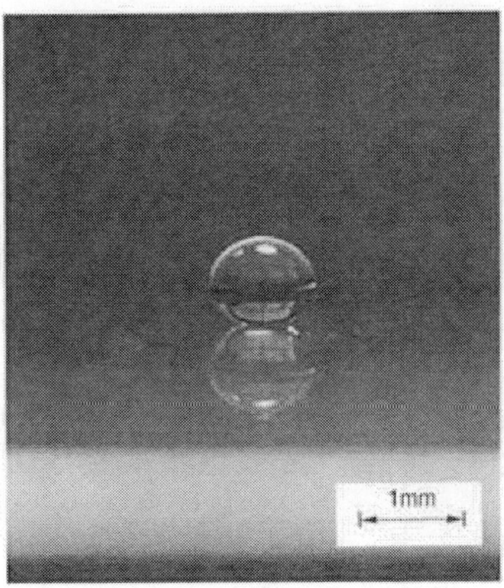

Figure 3. Photograph of a water droplet of about 1 mm in diameter on a glass plate with flower-like alumina thin film coated with hydrolyzed fluoroalkyl trimethoxysilanes.

Flower-like alumina films also show excellent antireflective (AR) properties in the wide visible light wavelength region on soda-lime glass, non-alkali glass and quartz glass substrates. The transmittance spectra of the coated substrates show high transparency of more than 99 % in visible light region, indicating that almost no light scattering occurs with the flower-like alumina. In the usual mono- or multi-layered AR films, the AR properties usually show large incident angle dependence. However, the AR effects based on the small roughness of the flower-like alumina show very small incident angle dependence.

3. Development and Application of Glass-Based Solid Electrolytes

3.1. Solid Electrolytes

In the previous section, it was demonstrated that a simple white powder of alumina could be changed to a variety of functional materials by structural, morphological, and compositional modifications. An additional example of a functional material is β-Alumina with extremely high ionic conductivity. High ionic conducting materials are called solid electrolytes. β-alumina, which does not have pure Al_2O_3 but a $Na_2O{\cdot}11Al_2O_3$ composition, exhibits a unique layered structure suitable to realize a fast Na^+ ion transport to the layers. This β-alumina is utilized as an electrolyte in sodium-sulfur batteries, in which Na metal and elemental S are used as electrode active materials. These batteries have attracted much attention as large-scaled batteries for power storage because of their very high energy density on the basis of the extremely large capacity of both Na and S.

In solids, constituent particles like ions, atoms, or molecules are packed together so that it is usually difficult for those particles to move quickly as opposed to the corresponding liquids. Fast ion conduction in solids thus require special design in the arrangement of the particles. Figure 4 shows typical crystal structures suitable to high ion conduction.

The first structure is a "defect structure" as in the case of so-called stabilized zirconia. If you add a small amount of CaO to ZrO_2, defects of O^{2-} generated as Zr^{4+} ions are replaced by Ca^{2+} ions. These O^{2-} defects lead to O^{2-} ion conduction in solids. The second structure is a "layered structure," typically seen in β-alumina. β-Alumina has a composition of $Na_2O{\cdot}11Al_2O_3$, as mentioned above, and a layered structure with weakly bonded inter-layers. The Na^+ ions, present at the inter-layers, easily migrate there. The last structure is an "average structure" with α-AgI as a typical example. In the structure of α-AgI, the arrangement of larger I^- ions is body-centered cubic so that there are two I^- ions in a unit cell. On the other hand, there are 42 candidate positions of Ag^+ ions with equal potential energy in the unit cell, as shown in figure 4, although the number of Ag^+ ions in the unit cell is only two. It was experimentally confirmed that most Ag ions were on average present in 12 tetrahedral positions, which explains the fast Ag^+ ion conduction in solids. α-AgI, which is a high temperature crystalline phase of AgI, exhibits the highest level of ion conductivity in all solids.

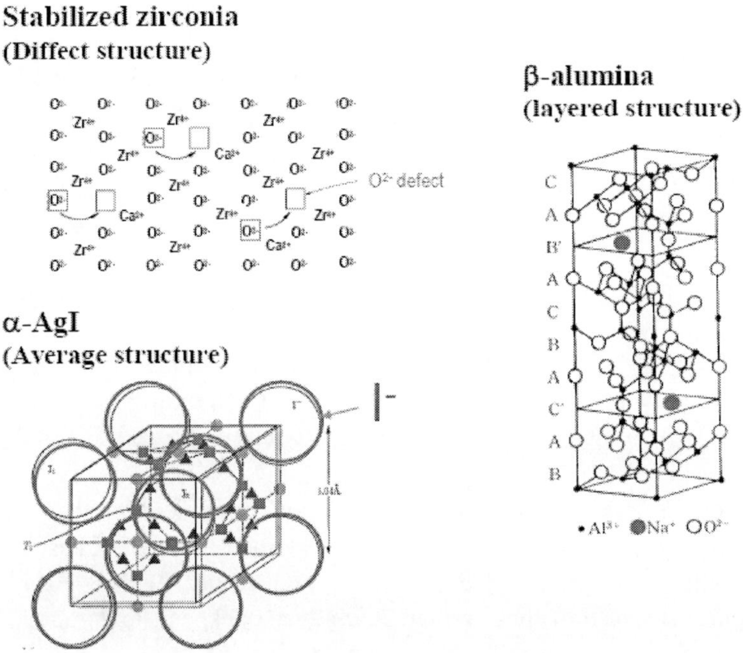

Figure 4. Typical crystal structures suitable to high ion conduction.

In the case of crystalline materials, the design of a special structure suitable to ion migration is essential to realize fast ion conduction, as seen in figure 4. We can take glassy materials which are another type of solid. Glass formation or vitrification is known as another way to realize fast ion conduction in solids. Glass with a random arrangement of constituent particles has a larger specific volume when compared to the corresponding crystal; using this excess volume, called free volume, ions can migrate more easily in glass structures. Higher ion conductivity is, therefore, usually observed more in glass than in corresponding crystal due to the open structure of glass. In addition, single ion conduction is realized because glassy materials belong to the so-called decoupled systems. Single ion conduction leads to an ideal battery system with no side reactions and the effective transport of carriers.

A series of oxyhalide glasses in the systems AgI-Ag_2O-M_xO_y (M=B, P, Ge, Mo, W, etc.), which show very high Ag^+ ion conductivities of 10^{-3} to 10^{-2} Scm^{-1} at room temperature, have been developed as one of the most ion-conductive glassy materials [2]. These are called superionic glasses. These oxyhalide glasses exhibit higher conductivity than corresponding crystals with the same compositions. In addition the Ag^+ ion transport number of these glasses is almost unity.

Stabilization of the superionic crystalline phases is another unique and remarkable advantage of glass-based electrolytes. Superionic crystalline phases, as a high-temperature phase or a metastable phase, are easily formed from glassy electrolytes. If you heat glass beyond the glass-transition temperature, crystallization usually occurs. In most cases, crystallization from glass results in the formation of a less conductive crystalline phase. However, if a much more conductive superionic crystalline phase, as mentioned above in figure 4, is present as high-temperature or a metastable phase, crystallization usually results in the formation of these phases. In fact, in the case of crystallization from AgI-based oxyhalide glasses as in an AgI-Ag_2O-MoO_3 system with an extremely high concentration of AgI, the high-temperature superionic phase of α-AgI, as shown in figure 4, is precipitated from the glasses and successfully stabilized in a glass matrix at room temperature [3]. The crystalline phase of α-AgI is usually stable at temperatures above 147°C and it easily transforms to less conductive β-AgI below the temperature; the room temperature conductivity is less than 10^{-7} Scm^{-1}. However, the α-AgI-stabilized materials obtained from AgI-based oxyhalide glasses exhibit extremely high Ag^+ ion conductivity of 10^{-1} Scm^{-1} at room temperature. Because of these remarkable advantages, as shown above, a variety of glass-based solid electrolytes have been developed.

3.2. Sulfide-Based Glass-Ceramic Electrolytes with Lithium Ion Conduction

Lithium ion conducting inorganic materials have recently been in the limelight as much safer electrolytes in lithium secondary batteries not only for portable electronic devices but also hybrid electric vehicles and all-electric vehicles. Figure 5 shows the temperature dependence for the conductivity of a variety of high lithium ion conductors including the crystalline and glassy materials reported so far.

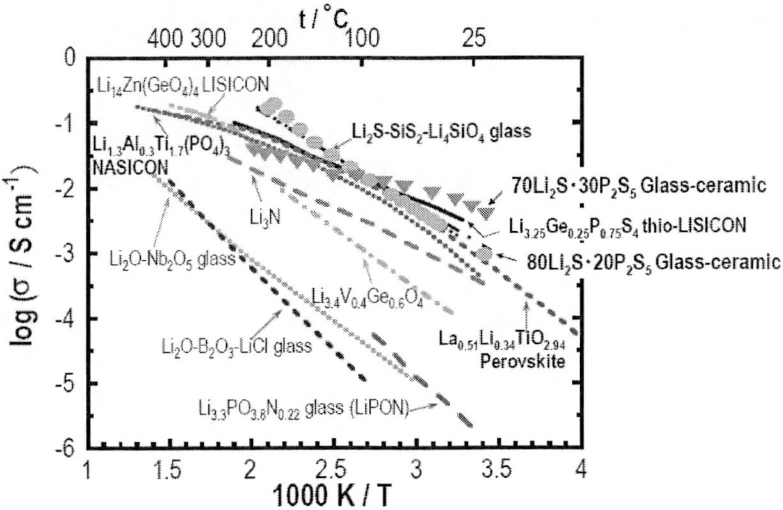

Figure 5. Temperature dependence of conductivity for a variety of high lithium ion conductors including the crystalline and glassy materials reported so far.

In the case of crystalline materials, NASICON-type $Li_{1.3}Al_{0.3}Ti_{1.7}(PO_4)_3$ and Perovskite $La_{0.51}Li_{0.34}TiO_{2.94}$ exhibit the same level of conductivity, and $Li_{3.25}Ge_{0.25}P_{0.75}S_4$ thio-LISICON shows the highest conductivity of 2.2 x 10^{-3} Scm^{-1} at room temperature [4]. “NASICON,” “Perovskite,” and “thio-LISICON” are names that represent the characteristics of a crystal structure. These crystalline electrolytes exhibit a designed structure suitable for fast Li^+ ion conduction, as seen in figure 4.

In glassy electrolytes, Li_2S-SiS_2-Li_4SiO_4 ($95(0.6Li_2S{\cdot}0.4SiS_2){\cdot}5Li_4SiO_4$) oxysulfide glass prepared by the twin-roller quenching of melts has the highest level of conductivity of more than 10^{-3} Scm^{-1} at room temperature. An ion conductivity of 10^{-4} Scm^{-1} is said to be the practical minimum for electrolytes in battery use. Several important guidelines for creating high Li^+ conducting glassy electrolytes have been proposed [5]. These include as much of an increase in the carrier concentration of Li^+ ions as possible, a change of the oxide matrix to a sulfide one, and the utilization of the so-called “mixed-anion effect” by combining sulfide and oxide anions. These guidelines are effective to increase Li^+ ion conductivity in glassy materials and they are used to advantage for high Li^+ ion conductivity in the Li_2S-SiS_2-Li_4SiO_4 oxysulfide glasses.

The conductivities of the $70Li_2S{\cdot}30P_2S_5$ and $80Li_2S{\cdot}20P_2S_5$ glass-ceramics, in which mechanochemically prepared glasses are crystallized, are $3.2 x 10^{-3}$ Scm^{-1} and $1.0 x 10^{-3}$ Scm^{-1}, respectively. Both exhibit the highest level of conductivity at room temperature of all the lithium ion conducting materials. Mechnochemical synthesis of glassy solid electrolytes has also recently been developed [6]. The mechanochemical process using a planetally ball mill is basically a room-temperature process where fine powders are obtained directly from the starting materials. Fine powders are very favorable in generating good solid-solid contact between the electrode and electrolyte materials in application to all-solid-state batteries.

Crystallization of glassy materials usually lowers the ionic conductivities. However, conductivity enhancement is observed by heat treatment of mechanochemically prepared Li_2S-P_2S_5 glasses. It is notable that after heat-treatment at each initial crystallization temperature of the Li_2S-P_2S_5 glasses, remarkable enhancement is observed at 70 and 80 mol%Li_2S compositions and different crystals are precipitated.

The X-ray diffraction patterns of the Li_2S-P_2S_5 glass-ceramics suggest that, in the case of $80Li_2S{\cdot}20P_2S_5$ (mol%) composition, a crystal analogous to the thio-LISICON II ($Li_{3.25}Ge_{0.25}P_{0.75}S_4$) phase, represented as $Li_{3+5x}P_{1-x}S_4$, is precipitated as the first phase after heat treatment of the $80Li_2S{\cdot}20P_2S_5$ glass. This leads to higher conductivity of the glass-ceramic than that of the original glass. On the other hand, in the case of the $70Li_2S{\cdot}30P_2S_5$ composition, a new crystalline phase $Li_7P_3S_{11}$, which cannot be obtained by conventional solid-state reaction, is precipitated. The structure of the glass-ceramic can recently be determined by using Raman spectroscopy and synchrotron X-ray powder diffraction data. The superionic metastable phases that are only precipitated from corresponding glasses are responsible for the high ionic conductivities above 10^{-3} Scm^{-1} of the $70Li_2S{\cdot}30P_2S_5$ and $80Li_2S{\cdot}20P_2S_5$ glass-ceramics. The formation of a superionic metastable phase is, thus, the most remarkable advantage of glass-based solid electrolytes.

4. All-Solid-State Batteries

4.1. All-Solid-State Lithium Secondary Battery Using Sulfide Solid Electrolytes

An all-solid-state lithium secondary battery system using inorganic solid electrolytes is one of the ultimate goals of rechargeable energy sources with high safety, reliability and energy density. Inorganic solid electrolytes are, thus, very promising for applications to all-solid-state batteries. Studies on all-solid-state lithium secondary batteries have proceeded in two directions: one is in the development of a thin-film battery and the other is a bulk-type battery. Figure 6 shows the schematic of thin-film and bulk-type all-solid-state batteries.

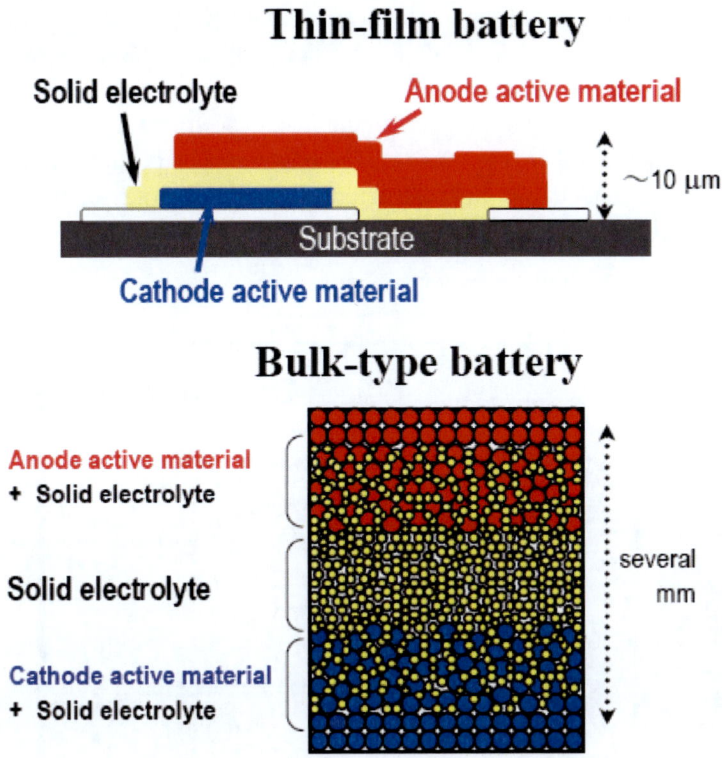

Figure 6. Schematic of thin-film and bulk-type all-solid-state batteries.

Bulk type all-solid-state batteries using glass-based sulfide solid electrolytes are basically constructed by a three-layered powder compressed pellet. The first layer is a negative electrode with anode active materials, e.g., Li metal-based alloys or carbon materials, the second layer is a solid electrolyte (SE), e.g., the Li_2S-P_2S_5 glass-ceramic powder, and the third layer is a positive electrode with cathode active materials, e.g., a composite with $LiCoO_2$, SE, and conductive additives like carbon nano-particles. The composite electrode is a mixture of three kinds of fine powders [7].

All-solid-state batteries using the $80Li_2S{\cdot}20P_2S_5$ glass-ceramic electrolytes as shown in the previous section and several active materials, such as $LiCoO_2$,

$LiNi_{1/2}Mn_{1/2}O_2$, V_2O_5, $Li_{4/3}Ti_{5/3}O_4$, C, etc., were fabricated. For example, the charge-discharge performances of all-solid-state cells of In / $80Li_2S{\cdot}20P_2S_5$ glass-ceramic / $LiCoO_2$ and In-Li / $80Li_2S{\cdot}20P_2S_5$ glass-ceramic / $Li_{4/3}Ti_{5/3}O_4$ are shown in figure 7. This figure shows the charge-discharge curves at the 100th, 200th, 500th and 700th cycles for all-solid-state cells using $LiCoO_2$ and $Li_{4/3}Ti_{5/3}O_4$ as an active material under 64 μAcm^{-2}. In spite of the 500th and 700th cycle data, high capacities and 100 % efficiency are still observed in the charge-discharge performance. These cells have an excellent cycle performance with no loss of capacity up to such a high cycle number. The advantage of glass-ceramics, in their high conductivity and dense microstructure, would promote a smooth charge-discharge reaction in the solid / solid interface between the electrolyte and electrode.

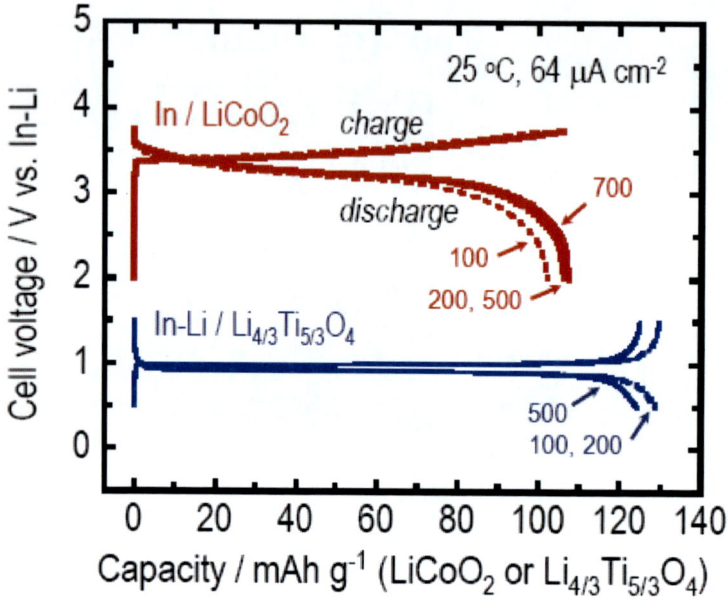

Figure 7. Charge-discharge performances of the all-solid-state cells of In / $80Li_2S{\cdot}20P_2S_5$ glass-ceramic / $LiCoO_2$ and In-Li / $80Li_2S{\cdot}20P_2S_5$ glass-ceramic / $Li_{4/3}Ti_{5/3}O_4$.

One problem with all-solid-state lithium secondary batteries is that it is difficult to use under very high electric current densities. This problem is probably due to the poor contact between solid electrolytes and solid electrodes in comparison to the solid-liquid contact in conventional batteries using liquid electrolytes because the contact in bulk-type all-solid-state batteries is just a point contact between solid particles in the electrode composites, as shown in figure 6. However, very recently, the coatings of $LiCoO_2$ particles with $LiNbO_3$ and $Li_5Ti_3O_8$ films have been reported to dramatically improve the rate performance of all-solid-state lithium secondary batteries using sulfide-based solid electrolytes. The coatings of $LiCoO_2$ particles with the CoS and Li_2O-SiO_2 film were also reported to improve the high performance rate of all-solid-state batteries. These findings will open the door in the development and application of bulk-type all-solid-state lithium secondary batteries for practical use.

4.2. All-Sulfide Batteries - A New Concept in All-Glass-Based Battery Systems

In order to improve the energy density of lithium secondary batteries, a variety of active materials with high electrochemical capacities have been widely investigated. Elemental sulfur is one of the most promising candidates of cathode active materials for next-generation secondary batteries because sulfur has a very large theoretical capacity of 1672 $mAhg^{-1}$, as mentioned in the previous section. In this case Li_2S is the discharged product. However, the Li / S battery system has a serious problem as a secondary battery; polysulfides represented as Li_2S_x formed in the discharge process are soluble in conventional liquid electrolytes so that Li / S batteries suffer from rapid fading of capacity upon cycling. The utilization of inorganic solid electrolytes, however, can resolve the problem since polysulfides are expected to be insoluble in solid electrolytes.

Due to the low electronic conductivity in pure S, all-solid-state Li / S batteries using S-Cu or S-C composites instead of pure S as a cathode active material were fabricated. The S-Cu composites, which are prepared by mechanical milling of elemental S and Cu, possess a core-shell structure with S particles covered with a CuS layer. The all-solid-state Li-In / $80Li_2S{\cdot}20P_2S_5$ glass-ceramic / S-Cu battery shows typical charge-discharge curves similar to the first cycle curves of the Li / S cell using liquid electrolytes. The charge-discharge performance tends to improve with an increase in the cycle number. Large capacities over 650 $mAhg^{-1}$ are observed even after the 20th cycle, which suggests that elemental S is utilized as active materials on charge-discharge processes in all-solid-state Li /S cells because the obtained capacity is almost twice as large as the theoretical capacity of CuS. Elemental S, which cannot be used with liquid electrolytes, is usable in all-solid-state batteries with the sulfide glass-ceramic electrolytes.

Li / S batteries using Li_2S-Cu composite cathode active materials were also fabricated and operated. In this case, cell operation can start from charging while the cell using a S-Cu composite cathode starts from discharging. The charge-discharge performance of the Li / S batteries using the Li_2S-Cu composite is very similar to that using the S-Cu composite.

High capacity anode active materials have also been studied for high energy density batteries. A series of glassy materials containing Sn as an anode active material have been developed. The SnS-P_2S_5 glasses, prepared from the mechanochemical reaction between SnS and P_2S_5, work as an anode active material in all-solid-state batteries, e.g. the $80SnS{\cdot}20P_2S_5$ glass / $80Li_2S{\cdot}20P_2S_5$ glass-ceramic / $LiCoO_2$ cell. It is noteworthy that a common network former P_2S_5 is used for the electrolyte and electrode in this battery system. From the charge-discharge curves of this cell, a large irreversible capacity is observed in the first cycle but charge and discharge capacities stay above 400 $mAhg^{-1}$ after the second cycle, which is much higher than in carbon anode materials. In the first charge process, the following two reactions (1) and (2) may occur and after the second discharge and charge processes the reversible reaction (2) may occur to form a Li-Sn alloy. It is also noteworthy that the Li_2S-P_2S_5 materials are formed around metallic tin (eq(1)), suggesting the self-formation of a high conductive $80Li_2S{\cdot}20P_2S_5$ electrolyte surrounding the anode active materials. In fact, a mixing of solid electrolyte powders to the anode active materials is not necessary, probably due to the self-formation of high ionic conductive sulfide electrolytes.

$$0.8SnS{\cdot}0.2P_2S_5 + 1.6Li^+ + 1.60e^- \rightarrow 0.8Li_2S{\cdot}0.2P_2S_5 + 0.8Sn \qquad (1)$$

$$Sn + xLi^+ + xe^- \leftrightarrow Li_xSn \qquad (2)$$

The success of this type of all-solid-state battery shown above will open the door to the fabrication of glassy monolithic cells, where a common network former is used for electrolyte and electrode materials. The glassy monolithic cell, which is expected to facilitate smooth solid-solid contact between the electrolyte and electrode, is very promising as a future all-solid-state battery [8].

Since sulfide electrolytes basically exhibit high performance, all-sulfide glass-based lithium secondary batteries of $80SnS{\cdot}20P_2S_5$ glass / $80Li_2S{\cdot}20P_2S_5$ glass-ceramic / $75Li_2S{\cdot}25Cu$ composite were fabricated. The mechanically milled Li_2S-Cu composite, Li_2S-P_2S_5 glass-ceramic, and mechanochemically prepared SnS-P_2S_5 glasses are used, respectively, as the cathode, electrolyte, and anode materials. Figure 8 shows a schematic of this type of sulfide-based all-solid-state lithium secondary batteries. The charge-discharge curves of the cell demonstrate a relatively high capacity of more than 500 mAh/g. The monolithic cell with a continuity for the lithium ion surrounding from the anode through the electrolyte to the cathode parts would reduce the interfacial charge transfer resistance between the electrode and electrolyte. The operation of all the sulfide batteries is a first step in the realization of an all-solid-state monolithic cell.

All-solid-state lithium secondary batteries using non-flammable solid electrolytes as a functional inorganic material are strongly desired because of their high safety, reliability, and energy density. We are approaching the ultimate goal of rechargeable energy devices by a variety of material innovations in science and technology.

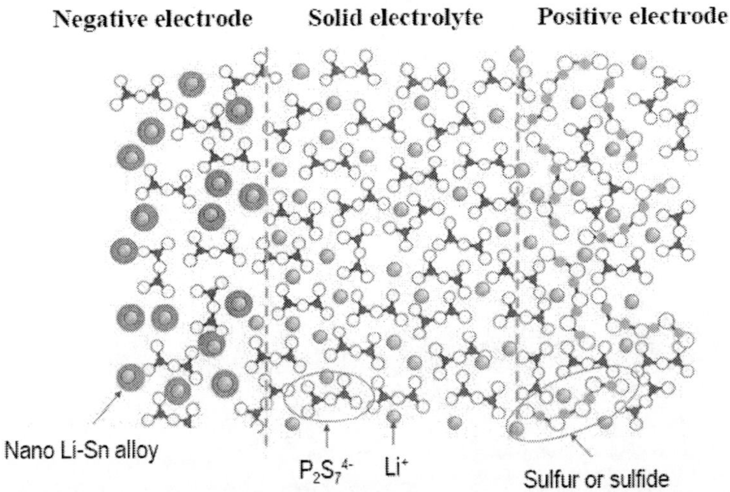

Figure 8. Schematic of monolithic cell using sulfide-based electrolytes and electrodes.

References

[1] K. Tadanaga, *J. Sol-Gel Sci. Tech.*, 40 (2006) 281.

[2] T. Minami, *J. Non-Cryst. Solids*, 95-96 (1987) 107.

[3] M. Tatsumisago, Y. Shinkuma and T. Minami, *Nature*, 354 (1991) 217; M. Tatsumisago, T. Saito and T. Minami, *Chem. Lett.*, (2001) 790.

[4] R. Kanno and M. Murayama, *J. Electrochem. Soc.*, 148 (2001) 742.

[5] T. Minami, A. Hayashi and M. Tatsumisago, *Solid State Ionics*, 177 (2006) 2715.

[6] M. Tatsumisago, *Solid State Ionics*, 175 (2004) 13.

[7] T. Minami, M. Tatsumisago, M. Wakihara, C. Iwakura, S. Kohjiya and I. Tanaka, *Solid State Ionics for Batteries*, Springer-Verlag Tokyo, 2005, p. 1.

[8] J.L. Souquet and M. Duclot, *Solid State Ionics*, 148 (2002) 375.

In: Environmentally Harmonious Chemistry... ISBN: 978-1-60876-428-0
Editors: M. Anpo and K. Mizuno

Chapter 3

Development of Environmentally-Friendly Artificial Photosynthetic Catalysts Using Clean and Abundant Solar Energy

Masaya Matsuoka and Masakazu Anpo

1. Introduction

Since the discovery of a Fe catalyst enabling ammonia synthesis from N_2 and H_2 by Fritz Haber in 1907, catalysis has played a crucial role in the development of modern science and technology. Many types of practical catalysts have been developed for the refinement of petroleum, polymer synthesis and the production of intermediate chemicals for industrial use. The majority of chemicals and fuels used in modern industry have, thus, been produced by advanced and sophisticated catalytic processes. However, rapid growth of chemical technologies based on mass production and consumption has caused serious environmental and health problems such as global air and water pollution, ozone depletion, climate change and energy shortages. The development of new and non-polluting catalytic processes which will contribute to a sustainable environment and can address global energy production needs is, thus, strongly and urgently desired. One of the most ideal catalytic processes is the so-called "artificial photosynthesis" phenomenon which realizes environmentally-friendly chemical processes using abundant and clean solar energy. As shown in figure 1, natural photosynthesis in plants enables one of the most significant "up-hill" reactions where photon energy is converted into chemical energy and stored in the bonds of glucose, accompanied by a large positive change in the Gibbs free energy ($\Delta G > 0$).

$$H_2O + CO_2 \xrightarrow{h\nu} 1/6(C_6H_{12}O_6) + O_2 \qquad \Delta G = 502 \text{ kJ/mol}$$

Since the world's first energy crisis in the early 1970's, much research has been devoted to the development of efficient systems that would enable the absorption and conversion of solar light into useful chemical energy. One of the most promising of such "artificial photosynthetic" reactions is the photocatalytic splitting of water to produce H_2 and O_2 under solar light.

$$H_2O \xrightarrow{h\nu} H_2 + 1/2\ O_2 \qquad \Delta G = 237 \text{ kJ/mol}$$

In 1972, Honda and Fujishima discovered the photosensitization effect of the TiO_2 electrode on the electrolysis of water into H_2 and O_2 [1,2]. Since this groundbreaking research, photocatalysis by TiO_2 semiconductors has been widely studied with the aim of converting light energy into reliable and effective chemical energy. The refinement of this up-hill reaction will be significant not only for the storage of solar energy but also for clean production of hydrogen since its consumption is expected to increase dramatically in coming generations, particularly for fuel cells [3-6]. In addition to the water decomposition reaction, artificial photosynthesis reaction systems, which can produce hydrocarbons and oxygen from carbon dioxide and water under solar light, has recently been developed with titanium oxide photocatalysts highly dispersed within the framework structure of zeolites as supports, as shown in figure 1 (b) [7-15].

It has also been shown that TiO_2 photocatalysts can efficiently induce "down-hill" reactions in addition to up-hill reactions. In a down-hill reaction, the photon energy absorbed by a photocatalyst induces thermodynamically favored reactions such as the complete oxidation of organic compounds into CO_2 and H_2O, accompanied by a large negative change in the Gibbs free energy ($\Delta G < 0$). Such reactions have been practically applied for the degradation of toxic organic compounds in air or water with solid semiconducting photocatalysts such as TiO_2 [2,16]. Moreover, recent developments in visible light-responsive TiO_2 photocatalysts which can react not only under UV but also natural light is of great importance in the realization of an effective artificial photosynthesis system [3-

10,17]. This chapter introduces the fundamental principles of photocatalysis, the development of visible light-responsive TiO_2 photocatalysts and their application for environmental protection and new energy production.

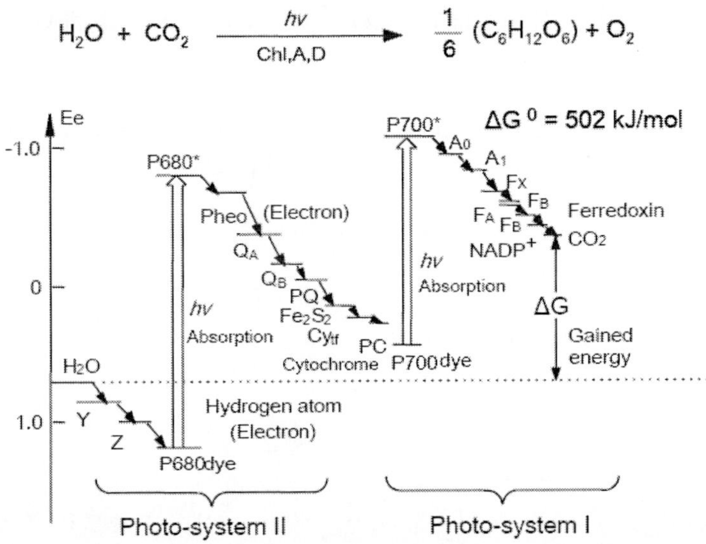

a

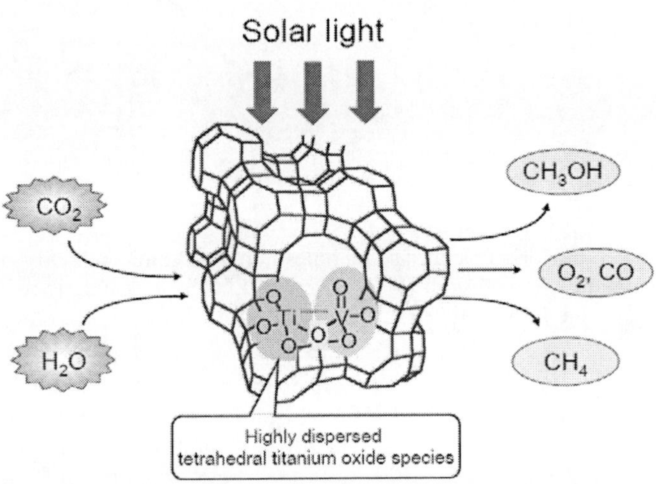

b

Figure 1. (a) Mechanism for photosynthesis in green plants; and (b) Artificial photosynthetic reactions on visible light-responsive titanium oxide photocatalyst incorporated within a zeolite (Ti/zeolite).

2. PRINCIPLES OF PHOTOCATALYSIS

Solar Energy Conversion and Environmental Applications

Semiconductors such as TiO_2, ZnO, Fe_2O_3 can act as sensitizers for light-induced redox processes due to their electronic structure which is characterized by a full valence band and empty conduction band [1-6]. As shown in figure 2, when the energy of a photon corresponds or exceeds the band gap energy of the semiconductor, an electron is promoted from the valence band into the conduction band, leaving a hole.

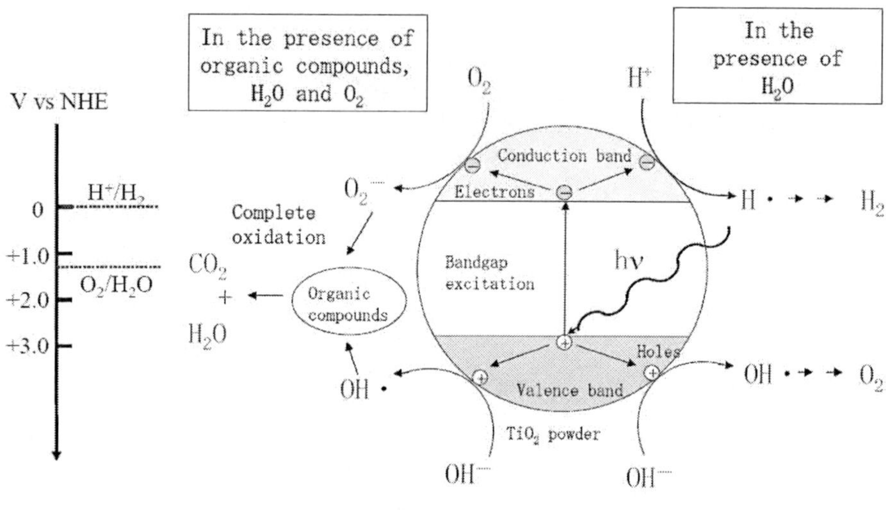

Figure 2. Photo-production of electrons and holes in n-type semiconductor TiO_2 particles: (Right) photocatalytic production of H_2 and O_2; and (Left) photocatalytic oxidation of organic compounds on TiO_2 particles suspended in water.

In the absence of suitable electron and hole scavengers (i.e., reactants), the stored electrons and holes are dissipated within a few nanoseconds by their recombination [1-6]. If a suitable scavenger (reactants) or surface defect state is available, recombination is prevented and redox reactions may subsequently occur by the remaining free electrons and holes. The valence band holes act as powerful oxidants while the conduction band electrons are good reductants [1-6]. When TiO_2 is irradiated by UV light ($\lambda < 380$ nm) in water in the absence of O_2, H^+ is reduced into H_2 by the photo-formed electrons while OH^- is oxidized into OH·

radicals and finally yields O_2 after several reaction steps. In this way, TiO_2 can decompose water into H_2 and O_2, enabling the efficient conversion of light energy into chemical energy. It has been noted that irradiation of UV light with shorter wavelengths than 165 nm (so-called vacuum ultraviolet) is necessary for the direct photolysis of water molecules into H_2 and O_2 [1-6]. On the other hand, when TiO_2 is irradiated by UV light in the presence of organic compounds together with water and air, the photo-formed electrons react with O_2 to form O_2^-, while OH^- is oxidized into OH· radicals. These oxygen radicals then easily react with the organic compounds, oxidatively decomposing them into CO_2 and H_2O. The application of such illuminated semiconductors in the remediation of environmental contaminants has been successful for a wide variety of compounds [2-6]. In many cases, the complete mineralization of the organic compounds has also been reported.

Among various photocatalysts, titanium oxide (TiO_2) is the most attractive and promising due to its low cost, availability, high photocatalytic reactivity and chemical stability. However, TiO_2 has a large band gap with an absorption edge in UV light regions shorter than 380 nm so that TiO_2 semiconductors are only able to absorb 3–4% of the solar light that reaches the earth. Thus, for indoor use, the use of special light sources such as a black light is necessary in activating the TiO_2 photocatalysts. In order to extend absorption of the photocatalysts into visible light regions, extensive investigations have been carried out and such attempts in the design of applicable visible light-responsive TiO_2 photocatalysts will also be introduced.

3. The Application of Photocatalysts for Solar Energy Conversion and Environmental Protection

3.1. Water Splitting Reaction to Produce Pure Hydrogen

Many studies have been devoted to water splitting reactions by various powdered photocatalysts for the production of H_2 as a clean fuel. For example, Pt-loaded TiO_2 has been applied for water splitting reactions under UV light irradiation. However, the reaction yields were low since the recombination reaction between H_2 and O_2 to produce H_2O occurs on the surface Pt particles which are additives to promote the water splitting reaction [1-2]. Moreover, a gas mixture of H_2 and O_2 usually evolves from powdered TiO_2 photocatalysts so that product separation is necessary for the practical use of H_2 as fuel. The development of a photocatalytic system enabling the separate evolution of H_2 and O_2 from water is, thus, strongly desired. Recently, an H-type reactor was constructed where the liquid phase is separated by a TiO_2 thin film device and proton exchange membrane for the separate evolution of H_2 and O_2 from water [3-6]. Furthermore, visible light-responsive TiO_2 thin films were specially prepared by a one-step magnetron sputtering deposition method by controlling the substrate temperature during the sputtering process [8-10]. As shown in figure 3, a visible light-responsive TiO_2 thin film device (Vis-TiO_2/Ti/Pt) consisting of a metal foil substrate deposited with a photocatalytic TiO_2 thin film on one side and Pt metal on the other by a magnetron sputtering deposition method was applied. Figure 4 shows the time profile of the separate evolution of H_2 and O_2 from water under solar light from a sunlight gathering system. It was clearly observed that sunlight irradiation of this TiO_2 thin film device within an H-type reactor successfully led to the stoichiometrical evolution of H_2 from the Pt and O_2 from the TiO_2 side [3-6]. Moreover, the rate of solar energy conversion efficiency reached 0.3 % for this reaction. Thus, these recent advances in photocatalytic processes have enabled the efficient conversion of solar energy into useful chemical energy, i.e., hydrogen energy, for new energy production processes.

In addition to the water splitting reaction, the photocatalytic evolution of H_2 from water containing biomass has also been the focus of much attention. Visible light irradiation of a photocatalytic TiO_2 thin film device immersed in water containing biomass led to the efficient evolution of H_2. Here, biomass as organic compounds were able to efficiently scavenge the photo-formed holes on the visible light-responsive TiO_2 thin film and enhance H^+ reduction into H_2 by the photo-formed electrons [3-6]. This reaction system produces CO_2 in addition to H_2 as a result of the oxidation of the biomass and the reduction of H_2O, however, it can be applied for a clean H_2 production system by combining it with an additional CO_2 system such as in agricultural systems which may efficiently promote photosynthesis by feeding the concentrated CO_2 to green plants.

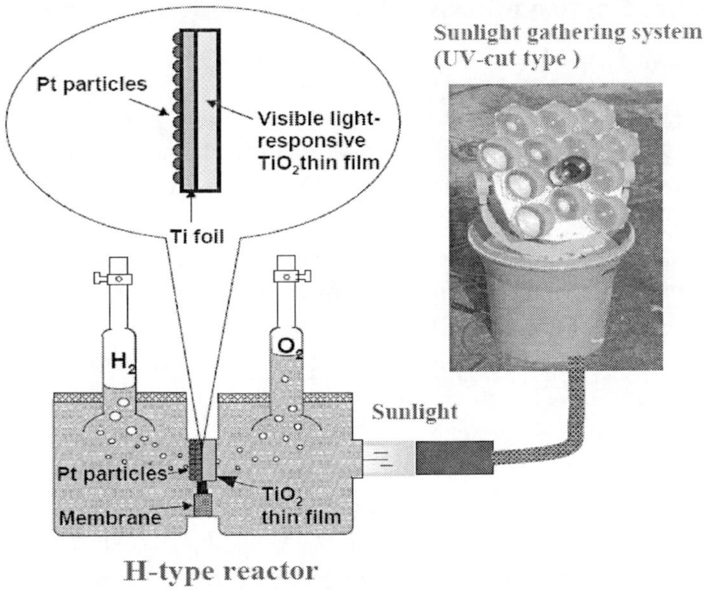

Figure 3. H-type reactor for the separate evolution of H_2 and O_2 using a visible light-responsive TiO_2 thin film device (Vis-TiO_2/Ti/Pt) under solar light irradiation. Electrolyte: 1.0 M NaOH aq (Right side); 0.5 M H_2SO_4 aq (Left side).

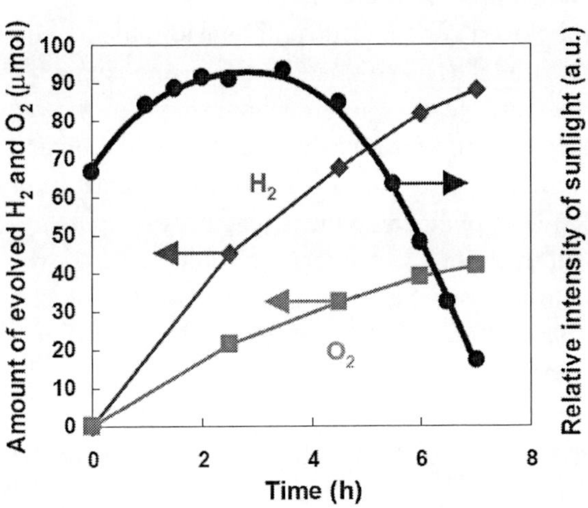

Figure 4. Separate evolution of H_2 and O_2 on a visible light-responsive TiO_2 thin film device (Vis-TiO_2/Ti/Pt) irradiated by light beams from a sunlight gathering system using an H-type reactor.

3.2. Artificial Photosynthesis: Photocatalytic Reduction of CO_2 with H_2O

The development of efficient photocatalytic systems which are able to reduce CO_2 with H_2O into chemically valuable compounds such as CH_3OH or CH_4 under solar light is an especially challenging goal in the search for environmentally-friendly catalysts. This is a typical up-hill reaction which converts solar energy into chemical energy as an artificial photosynthetic reaction. Up to now, the gas phase reaction of CO_2 and H_2O on powdered TiO_2 photocatalysts has yielded CH_4 as the major product, accompanied by the formation of CO, C_2H_6 and C_2H_4 as minor products [11-15]. Moreover, product distributions were affected by the kinds of co-catalysts used such as Pt or Cu, i.e., Pt-TiO_2 and Cu-TiO_2 yielded CH_4 and CH_3OH as the major product, respectively.

Recently, highly dispersed tetrahedrally-coordinated titanium oxide species within zeolite frameworks have been shown to exhibit high and unique photocatalytic reactivity for the reduction of CO_2 with H_2O when compared with bulk semiconducting powdered TiO_2 catalysts [11-15]. The electrons and holes produced in the conduction and valence bands, respectively, of the semiconducting TiO_2 powdered catalysts under UV light irradiation have been found to play a major role in various photocatalytic reactions. However, with a decrease in the particle size of the TiO_2 catalyst to less than 100 Å, a higher efficiency in the reactions could be observed [11]. As the size of the TiO_2 particles is reduced below a certain critical dimension, the gap between its highest occupied molecular orbital (HOMO) and lowest unoccupied molecular orbital (LUMO) starts to increase, leading to an enhancement of the reduction ability of the photo-formed electrons in LUMO as well as the oxidation ability of the photo-formed holes in HOMO (figure 5) [11]. This "size quantization effect" leads to high and selective photocatalytic reactivity quite different from photoelectrochemical reactions occurring on bulk TiO_2 powders [11] and is attributed to an electronic modification of the TiO_2 catalysts as well as the close existence of the photo-formed electron and hole pairs and their balanced contribution to the reactions.

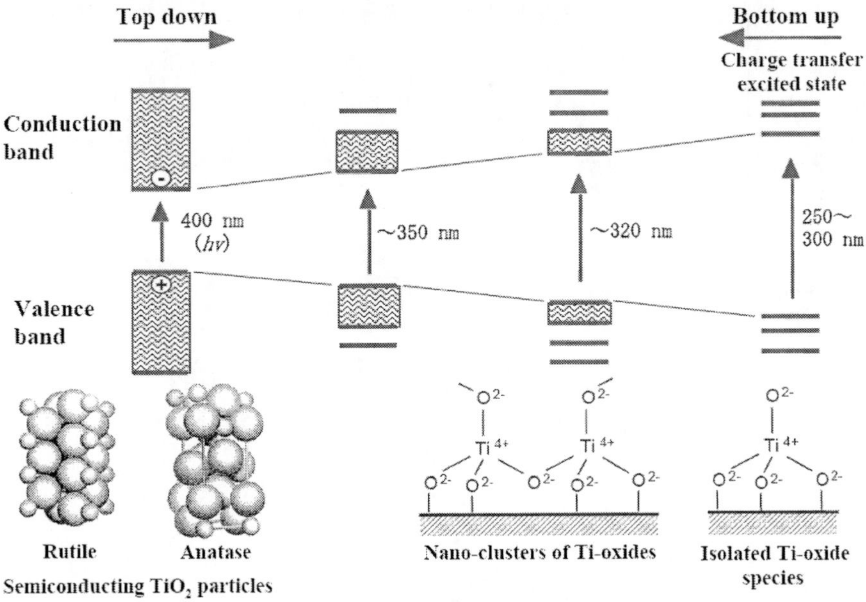

Figure 5. Progress of titanium oxide photocatalysts from extended semiconducting TiO_2 particles, nano-scale molecular clusters to an isolated titanium oxide species as well as the changes in their electronic states.

As shown in figure 1 (b), the titanium oxide species incorporated within zeolite frameworks (Ti/zeolites) exist in an isolated four-fold coordination sphere with a Ti–O bond distance of about 1.83 Å [11-15]. These Ti-containing zeolite catalysts exhibited a photoluminescence spectrum at around 480–490 nm by excitation at around 220–260 nm (figure 6). This photoluminescence spectrum is attributed to the radiative decay process from the charge-transfer excited state to ground state of the highly dispersed titanium oxide species in tetrahedral coordination, expressed as follows [11-15]:

$$(Ti^{4+} - O^{2-}) \underset{h\nu'}{\overset{h\nu}{\rightleftarrows}} (Ti^{3+} - O^{-})^{*}$$

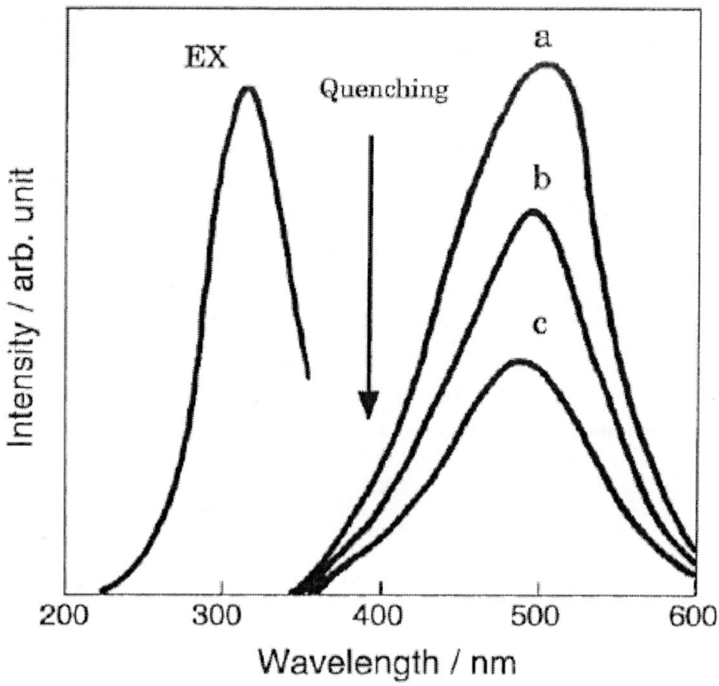

Figure 6. Ordinary photoluminescence spectrum observed for: (a) Ti/Y-zeolite catalyst prepared by an ion-exchange method; (EX) its corresponding excitation spectrum; (b and c) the effects of the addition of CO_2 and H_2O, respectively, at 77 K. Excitation at 290 nm; Emission monitored at 490 nm. Amount of added CO_2: (b) 8.5 mmol/g; H_2O: (c) 2.9 mmol/g.

The addition of H_2O or CO_2 molecules onto the Ti/zeolite catalysts led to an efficient quenching of the photoluminescence as well as shortening of the lifetime of the charge transfer excited state, its extent depending on the amount of gases added. Such efficient quenching of the photoluminescence with the H_2O or CO_2 molecules suggests not only that the four-fold coordinated titanium oxide species locate at positions accessible to these small molecules but also that they interact with these titanium oxides in both their ground and excited states. In fact, UV light irradiation of the Ti/zeolite catalysts in the presence of a mixture of CO_2 and H_2O led to the photocatalytic reduction of CO_2 to form CH_3OH and CH_4 as major products as well as CO, O_2, C_2H_4, and C_2H_6 as minor products in the gas phase at 323 K, while the yields of these photoformed products increased with a good linearity against the irradiation time [11-15]. These results clearly indicate that highly dispersed tetrahedrally-coordinated titanium oxide species exhibit high selectivity as well as efficiency for CH_3OH formation in the photocatalytic reduction of CO_2 with H_2O as compared with bulk TiO_2 with an octahedral coordination. Figure 7 shows the relationship between the coordination number of the titanium oxide species obtained from XAFS analysis and the selectivity for the formation of CH_3OH in the photocatalytic reduction of CO_2 with H_2O on various Ti/zeolite catalysts. A clear dependence of the selectivity for the formation of CH_3OH on the coordination numbers of the titanium oxide species can be observed; i.e., the lower the coordination number of the titanium oxide species, the higher the selectivity for CH_3OH formation. Bulk TiO_2 semiconducting photocatalysts did not show any reactivity for the formation of CH_3OH from CO_2 and H_2O. From these results, it could be proposed that the highly efficient and selective photocatalytic reduction of CO_2 into CH_3OH by H_2O was, in fact, achieved with Ti/zeolite catalysts involving highly dispersed four-fold coordinated titanium oxides in their frameworks as the active species [11-15].

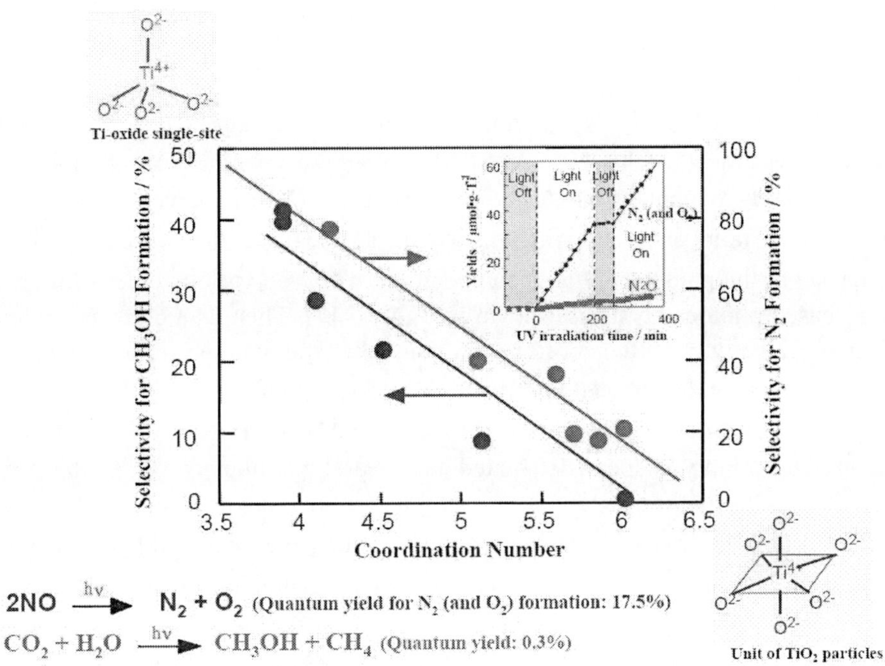

Figure 7. Relationship between the coordination numbers and photocatalytic reactivity of the titanium oxides.

3.3. Direct Photocatalytic Decomposition of NO into N_2 and O_2

The development of efficient photocatalytic systems which can decompose NOx directly into N_2 and O_2 is essential in establishing clean and sustainable deNOx systems for atmospheric purification. It has been reported that the decomposition of NO can proceed photocatalytically on powdered TiO_2 at room temperature and N_2O is produced as the major product [11-14,18]. In order to decompose NO directly into N_2 and O_2, highly dispersed titanium oxide photocatalysts were prepared and their photocatalytic activities were investigated. UV light irradiation of the powdered TiO_2 and Ti/Y-zeolite catalysts prepared by ion-exchange (ex-Ti/Y-zeolite) or impregnation (imp-Ti/Y-zeolite) methods in the presence of NO led to the evolution of N_2, O_2 and N_2O in the gas phase at 275 K with differing yields and product selectivity [11-14,18]. The yields of the photo-formed products increased linearly against the UV irradiation time and the reaction immediately ceased when irradiation was discontinued. A comparison of the photocatalytic activity of the Ti/Y-zeolite catalysts and widely used bulk TiO_2 powdered catalyst showed that specific photocatalytic reactivity of the Ti/Y-zeolite catalysts, which was normalized for the unit amount of TiO_2 in the catalysts, was much higher [11-14,18]. Moreover, the selectivity for the formation of N_2 strongly depended on the type of catalyst. The relationship between the coordination number of the titanium oxide species and the selectivity for N_2 formation in the photocatalytic decomposition of NO on various types of titanium oxide-based photocatalysts are shown in figure 7 [7]. There is a clear dependence of the N_2 selectivity on the coordination number of the titanium oxide species [11-14,18]. These results showed that the highly efficient and selective photocatalytic reduction of NO into N_2 and O_2 could be achieved with the ex-Ti/Y-zeolite which incorporates highly dispersed isolated tetrahedral titanium oxide as the active species. The formation of N_2O as the major product was also observed for bulk TiO_2 and imp-Ti/Y-zeolite catalysts, which include the octahedrally-coordinated aggregated titanium oxide species. Based on these results, the reaction mechanism for the photocatalytic decomposition of NO on the isolated tetrahedral titanium oxide species could be proposed, as shown in Scheme 1. NO molecules could adsorb onto the oxide species as weak ligands to form the reaction precursors. Under UV light irradiation, the charge-transfer excited complexes of the oxides, $(Ti^{3+}—O^{-})^*$, were formed. Within their lifetimes, the electron transfer from the trapped electron center, Ti^{3+}, into the π-antibonding orbital of NO takes place and, simultaneously, the electron transfer from the π-bonding orbital of another

NO into the trapped hole center, O^-, occurs. These electron transfers led to the direct decomposition of two sets of NO on $(Ti^{3+}—O^-)^*$ into N_2 and O_2 under UV irradiation in the presence of NO, even at 275 K. With the aggregated or bulk TiO_2 catalysts, the photo-formed holes and electrons rapidly separate spatially from each other, with large distances between the holes and electrons, thus, preventing the simultaneous activation of two NO on the same active sites and resulting in the formation of N_2O and NO_2 in place of N_2 and O_2 [11-14,18]. Moreover, the decomposed N and O species react with NO on different sites to form N_2O and NO_2, respectively. It was, thus, clearly demonstrated that the use of zeolites as supports could enable the anchoring of a titanium oxide species in a highly dispersed state within the zeolite cavities and such tetrahedrally coordinated titanium oxide photocatalysts were found to be promising candidates in removal systems for toxic NOx compounds in the atmosphere.

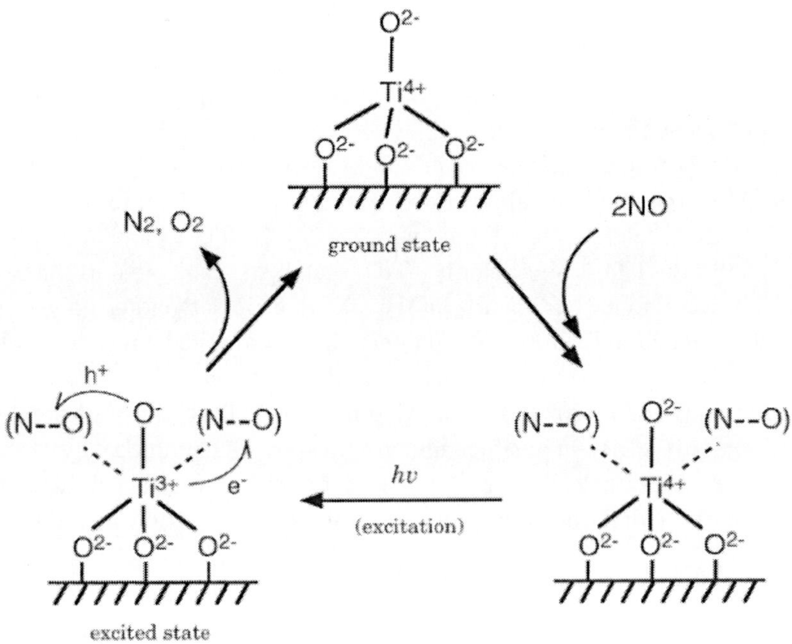

Scheme 1. Reaction scheme of the photocatalytic decomposition into N_2 and O_2 on the ex-Ti/Y-zeolite catalyst at 275 K.

3.4. Applications in the Purification of Water Polluted with Various Toxic Compounds Such as Dioxins

When TiO_2 is irradiated by UV light in the presence of organic compounds together with water and air, the photo-formed electrons react with O_2 to form O_2^-, while OH^- is oxidized into OH· radicals [2,16]. The formed oxygen radicals easily react with the organic compounds and decompose them into CO_2 and H_2O. In fact, this strong oxidation ability of the TiO_2 photocatalysts has been applied to the purification of water polluted with various types of toxic compounds such as dioxins which were completely mineralized into CO_2, H_2O and HCl under UV light irradiation [2,16]:

$$C_nH_mO_zCl_y \xrightarrow[h\nu]{O_2} nCO_2 + yHCl + wH_2O$$

These TiO_2 photocatalysts have been practically applied for the purification of underground water polluted with volatile organic compounds (VOC) such as tetrachloroethylene [2,16].

It has also been reported that visible light-responsive TiO_2 photocatalyts can be prepared by applying a metal ion-implantation method. Metal ions such as Fe^+, Mn^+, and V^+ are accelerated to a high kinetic energy (150 keV) and implanted into the deep bulk of TiO_2. Signficantly, TiO_2 catalysts subjected to metal ion-implantation were found to absorb and work as efficient photocatalysts under visible light irradiation. These TiO_2 photocatalysts exhibited high reactivity for the liquid-phase degradation of 2-propanol diluted in water at 295 K under visible light irradiation (λ>450 nm), as shown in figure 8 [19]. This advanced preparation method for highly efficient visible light responsive TiO_2 photocatalysts can be considered a breakthrough in the widespread practical application of photocatalysis in addressing large scale environmental pollution including water purification.

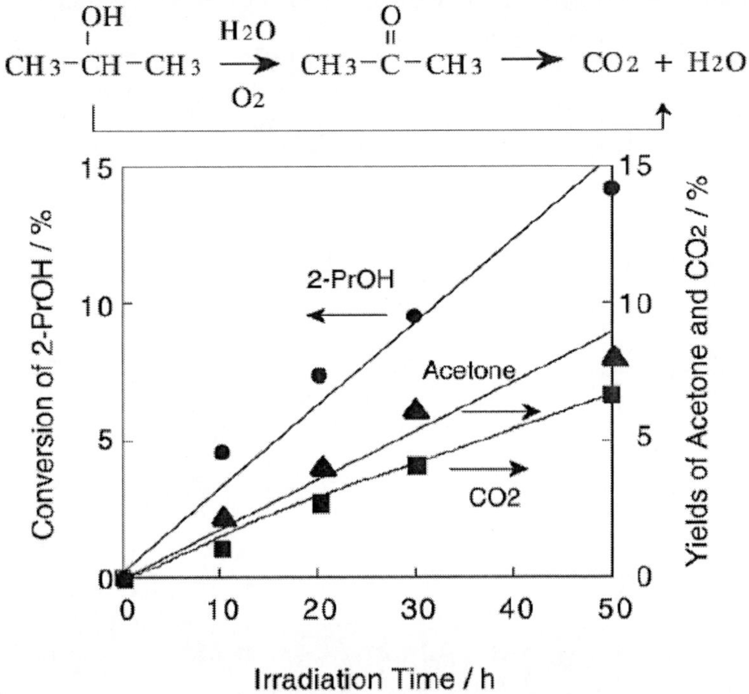

Figure 8. Reaction time profiles of the photocatalytic oxidative degradation of 2-propanol diluted in water on the Fe ion-implanted TiO_2 photocatalysts under visible light irradiation ($\lambda > 450$ nm). Ion acceleration energy: 150 keV. Amount of implanted Fe ions: 13.2×10^{-7} mol/g-cat.

3.5. Superhydrophilicity of TiO_2 Thin Films and Their Application in Self-Cleaning Materials

Various studies have also led to TiO_2 thin films which exhibit superhydrophilicity under UV light irradiation [2]. As shown in figure 9, the contact angle of water droplets on a TiO_2 thin film reaches almost zero under UV light, thus, achieving a superhydrophilic state. Under such a superhydrophilic state, water can spread perfectly across the TiO_2 surface. Such superhydrophilic TiO_2 thin films have been applied to anti-fogging mirrors on which water cannot form droplets. Furthermore, this superhydrophilicity has also been applied to provide self-cleaning properties on such materials as window glass or architectural tiles since the high wettability of the TiO_2 thin film surface can prevent the adhesion of oily grime or dirt on the TiO_2 surface.

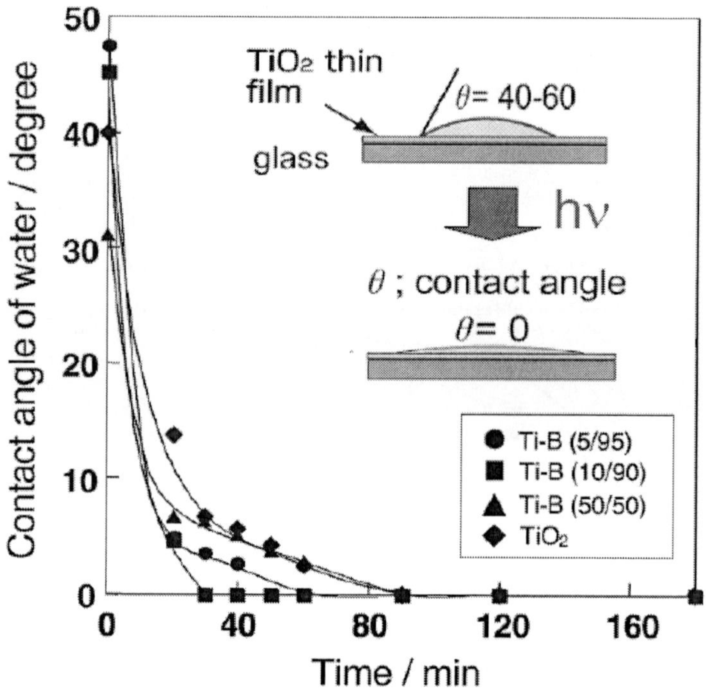

Figure 9. Changes in the contact angle of water droplets under UV light irradiation on Ti/B binary oxide thin films.

Ti/B binary oxide thin films including ultrafine TiO_2 nano-particles in octahedral coordination and dispersed in a host B_2O_3 compound have also been successfully prepared by an ionized cluster beam (ICB) deposition method using multi-ion sources [10,20]. As shown in figure 9, Ti/B binary oxide thin films prepared by ICB deposition were found to exhibit higher photo-induced hydrophilic properties as compared to the pure and untreated TiO_2 thin films [10,20]. These advanced binary oxide thin films can also be applied for efficient self-cleaning materials that can realize a cleaner and more toxic-free living environment.

4. Design of Visible Light-Responsive TiO_2 Photocatalysts

4.1. Modification of the Electronic State of TiO_2 by Applying an Advanced Metal Ion-Implantation Method

TiO_2 semiconductors have a relatively large band gap of 3.2 eV, corresponding to wavelengths shorter than 388 nm, thus, making use of only 3–4 % of the solar beams that reach the earth. However, the design of TiO_2 photocatalysts which can operate efficiently under both UV and visible light irradiation would be ideal for practical and widespread use, particularly with respect to environmental remediation. We have, thus, focused our attention on the development of visible light responsive TiO_2 photocatalysts by applying a metal ion-implantation method to modify the electronic properties of bulk TiO_2 photocatalysts by bombarding them with high-energy metal ions [8-10]. In our studies, metal-ion implantation with various transition-metal ions such as V, Cr, Mn, Fe, and Ni was found to cause a large shift in the absorption band of the TiO_2 catalysts toward visible light regions with differing levels of effectiveness. As can be seen in figure 10, the absorption band of Cr ion-implanted TiO_2 shifts smoothly towards the visible region, the extent depending on the amount of metal ions implanted [8-10]. Meanwhile, the absorption spectra of the TiO_2 catalysts impregnated or chemically doped with Cr ions were found to show no shift in the absorption edge of TiO_2, although a new absorption band appears at around 420 nm as a shoulder peak due to the formation of an impurity energy level within the band gap. These results indicate that the method of doping causes the electronic properties of the TiO_2 catalyst to be modified in completely different ways and also confirms that only advanced metal ion-implanted TiO_2 catalysts show such shifts in the absorption band towards visible regions.

The photocatalytic activity of the Cr ion-implanted TiO_2 was examined for the direct decomposition reaction of NO. As shown in figure 11, the unimplanted original and chemically doped TiO_2 catalysts show no activity for the decomposition of NO under visible light irradiation ($\lambda > 450$ nm) [8-10]. However, visible light irradiation ($\lambda > 450$ nm) of the Cr ion-implanted TiO_2 catalysts led to the decomposition of NO into N_2, O_2, and N_2O with a good linearity against the irradiation time. Metal ion-implanted TiO_2 catalysts were, thus, found to enable the absorption of visible light up to wavelengths of 400–600 nm and were also able to operate effectively as photocatalysts.

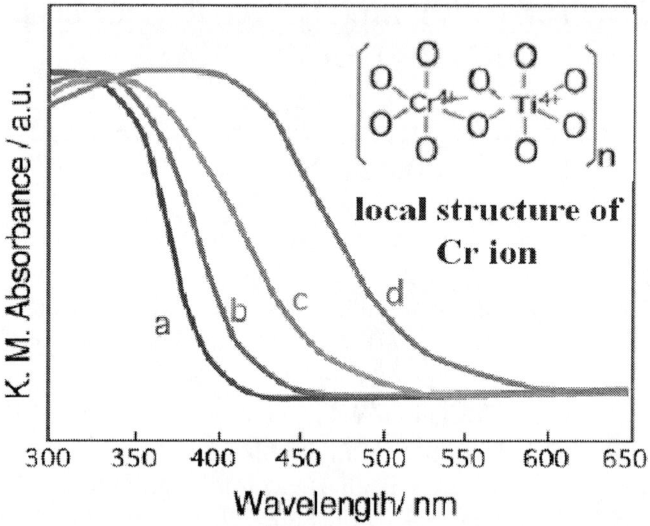

Figure 10. UV–Vis absorption spectra of: (a) TiO_2; and (b–d) Cr ion-implanted TiO_2 photocatalysts. Amount of implanted Cr ions (μmol/g): (a) 0, (b) 0.22, (c) 0.66, (d) 1.3.

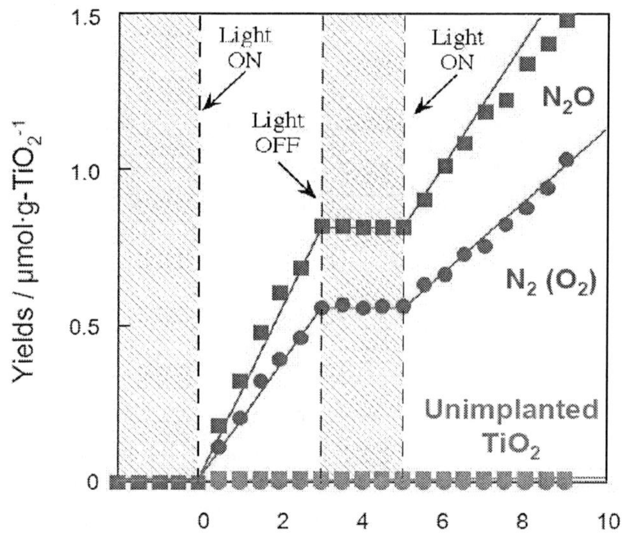

Figure 11. Time profiles of the direct photocatalytic decomposition of NO into N_2 and N_2O on the unimplanted pure TiO_2 catalyst and the Cr ion-implanted TiO_2 catalyst with Cr ions of $6.6x10^{-7}$ mol/g-TiO_2 under visible light irradiation ($\lambda > 450$ nm).

It was noted that the photocatalytic reactivity of the metal ion-implanted TiO_2 catalysts retained the same photocatalytic efficiency as the unimplanted original TiO_2 catalyst under UV light (λ < 380 nm). When metal ions were chemically doped into the TiO_2 catalyst, the photocatalytic efficiency decreased dramatically under UV light irradiation due to the quick recombination of the photoformed electrons and holes through the impurity energy levels formed by the doped metal ions within the bandgap of the catalyst. The physically implanted metal ions, thus, did not work as electron–hole recombination centers but only induced a modification of the electronic properties of the catalyst.

The local environment of the implanted metal ions was investigated by XAFS (XANES and EXAFS) measurements [8-10]. XAFS analysis showed that, in the TiO_2 catalysts chemically doped with Cr ions by an impregnation method, the ions were present as aggregated Cr oxides having an octahedral coordination similar to that of Cr_2O_3 and a tetrahedral coordination similar to that of a mixture of Cr_2O_3 and CrO_3. On the other hand, in the catalysts physically implanted with Cr ions, the ions were present in a highly dispersed and isolated state in octahedral coordination, clearly indicating that the Cr ions were incorporated into the lattice positions of the TiO_2 catalyst in place of the Ti ions. These findings show that a modification of the electronic state by metal ion-implantation is closely associated with the strong and long-distance interactions which arise between the TiO_2 and metal ions implanted and not with the formation of impurity energy levels within the bandgap of the catalysts.

4.2. Development of Visible Light-Responsive Ti/Zeolite Catalysts by an Advanced Metal Ion Implantation Method

Titanium oxide photocatalysts anchored within various zeolites exhibited unique and high photocatalytic activity for various reactions such as the direct decomposition of NO into N_2 and O_2 or the reduction of CO_2 with H_2O. However, the isolated tetrahedral Ti^{4+} oxide species absorbs UV light of wavelengths below 300 nm since the HOMO-LUMO energy gap of this isolated tetrahedral Ti^{4+} oxide species becomes significantly larger than that of bulk TiO_2 due to the size quantization effect. In other words, titanium oxide photocatalysts cannot utilize the abundant solar energy that reaches the earth, necessitating a UV light source for its use as a photocatalyst. Photocatalysts which can operate efficiently under both UV and visible light irradiation are, thus, essential in harnessing the vast power and possibilities of safe and clean solar energy. A metal ion-implantation method has been applied to modify the electronic properties of the Ti/zeolite photocatalyst by bombarding them with high-energy metal ions, leading to the discovery that various transition metal ions such as V, Cr, accelerated by high electric fields can produce a large shift in the absorption band of Ti/zeolite photocatalysts toward visible light regions [8-12]. Figure 12 shows the effect of V-ion implantation on the diffuse reflectance UV–Vis absorption spectra of Ti-containing mesoporous materials, Ti-HMS and Ti-MCM-41. Their absorption spectra at around 200–260 nm can be attributed to the charge-transfer absorption process involving an electron transfer from the O^{2-} to Ti^{4+} ion of the highly dispersed tetrahedrally-coordinated titanium oxide species of these catalysts [8-12]. The spectra were observed to shift smoothly toward visible light regions, the extent strongly depending on the amount of V ions implanted. The interaction of the implanted V ions with the tetrahedrally-coordinated titanium oxide species were, thus, seen to lead to a modification of the electronic properties of the titanium oxide species within the zeolite framework. The V K-edge FT-EXAFS spectra of the Ti-HMS catalyst implanted with V ions show that the next-neighbors of the V environment are not the same as for vanadium-oxide based catalysts (e.g., V_2O_5) and suggest the formation of tetrahedral titanium oxides having a V–O–Ti bond instead of V–O–V linkages (figure 1(b)) [8-12]. These findings indicate that the formation of the V–O–Ti bridge structures between the isolated tetrahedrally coordinated titanium oxide species and implanted V ions affect the electronic structure of the isolated titanium oxide species, leading to a red shift in the absorption spectra of these catalysts.

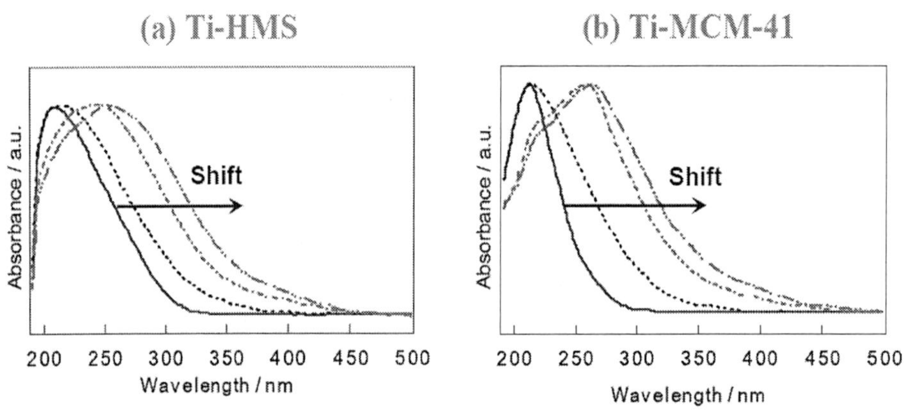

Figure 12. UV-Vis absorption spectra of the V ion implanted: (a) Ti-HMS; and (b) Ti-MCM-41 catalysts. Implanted-V ions (from left to right): 0, 0.66, 1.3, 2.0 (μmol/g-cat).

The photocatalytic activity of the V ion-implanted Ti-HMS and Ti-MCM-41 was investigated for the decomposition of NO into N_2 and O_2 under visible light (λ> 420 nm). Irradiation of the V ion-implanted Ti-HMS led to the efficient decomposition of NO into N_2 and O_2, while the unimplanted original Ti-HMS exhibited no activity for the reaction even under the same reaction conditions [8-12]. Moreover, no NO decomposition could be confirmed under UV (λ < 300 nm) or visible light (λ > 420 nm) on the V ion-implanted HMS. These results show that ion-implantation is an effective technique for the modification of the electronic properties of titanium oxide photocatalysts, enabling them to absorb visible light (λ >420 nm) and operate as highly efficient photocatalysts.

4.3. Preparation of Visible Light-Responsive TiO_2 Thin Film Photocatalysts by a RF Magnetron Sputtering Deposition Method

In our recent work, a simple one-step preparation process for visible light-responsive TiO_2 thin films has been successfully developed by applying a RF magnetron sputtering (RF-MS) deposition method, as shown in figure 13 [3-6,8-10]. The system consists of a substrate (quartz or Ti foil) equipped at the center position in the chamber parallel with the source material of the calcined TiO_2 plate. The calcined plate is sputtered by Ar plasma by inducing a RF power of 300 W in Ar atmosphere and the TiO_2 thin film is prepared on the quartz or Ti metal foil substrate mounted on a heater. The substrate temperature (T_S) was kept at a fixed degree between the range of 473 - 873 K. Figure 14 shows the effect of the T_S on the UV-Vis transmission spectra of the thin films prepared on a quartz substrate with a TiO_2 thickness of 1.2 μm. TiO_2 thin films prepared at 473 K (UV-TiO_2) showed no absorption in wavelengths longer than 380 nm, while the films prepared at a T_S higher than 673 K (Vis-TiO_2) were yellow-colored and exhibited considerable absorption in wavelengths longer than 380 nm, enabling the absorption of visible light [3-6,8-10]. Moreover, the edge of the absorption band of Vis-TiO_2 prepared at 873 K (Vis-TiO_2-873) was found to shift toward longer wavelength regions of around 600 nm as compared to around 400 nm for UV-TiO_2 prepared at 473 K. It was, thus, found that precise control of the substrate temperature enabled the development of unique visible light-responsive TiO_2 thin film photocatalysts and the efficiency of visible light absorption was observed to increase with an increase in the substrate temperature. It was also noted that the coexistance of O_2 with Ar in the sputtering chamber led to the formation of UV light-responsive TiO_2 thin films, regardless of the substrate temperature (Ts > 473 K) during the sputtering deposition process. These findings suggest that TiO_2

deposition under pure Ar gas without any trace of O_2 is one of the major factors in realizing the preparation of Vis-TiO_2 thin film photocatalysts. The prepared Vis-TiO_2 thin films were then found to act as efficient photocatalysts for such significant reactions as the separate evolution of H_2 and O_2 under solar or visible light irradiation (figure 4.) as well as the complete oxidation of organic compounds into CO_2 and H_2O.

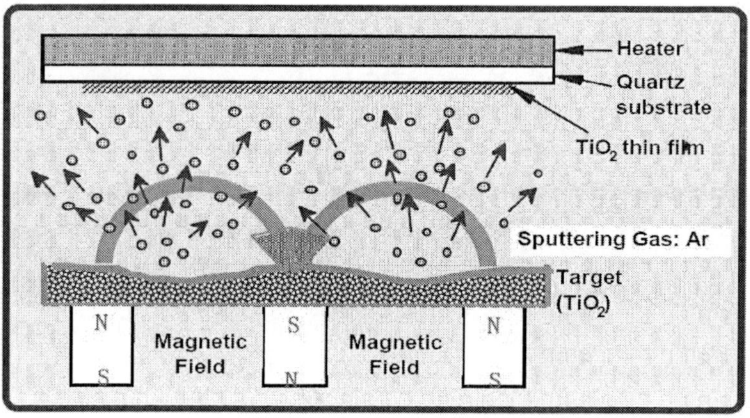

Figure 13. Schematic diagram of the RF magnetron sputtering (RF-MS) deposition method.

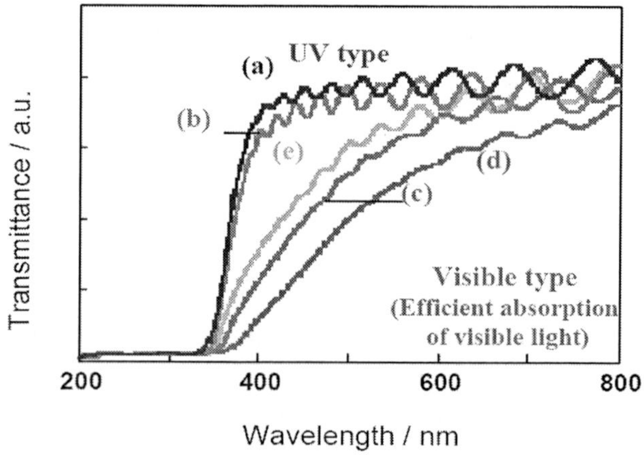

Figure 14. UV-Vis absorption (transmittance) spectra of TiO_2 thin films prepared on a quartz substrate by the RF-MS deposition method. Preparation temperatures (K): (a) 373, (b) 473, (c) 673, (d) 873, (e) 973.

To clarify the origin of visible light absorption for these Vis-TiO_2 thin films, SIMS and TEM analyses were carried out. The secondary ion intensity due to ^{18}O for UV-TiO_2 exhibited a stoichiometric O/Ti value of 2.00, independent of the depth from the TiO_2 surface, while in the case of Vis-TiO_2 , the O/Ti value gradually decreased from the top surface (O/Ti ratio: 2.00 ± 0.01) to the inside bulk (1.93 ± 0.01), showing a distinct contrast to UV-TiO_2 [3-6,8-10]. It was, thus, suggested that the unique declined composition of Vis-TiO_2 thin films of anisotropic structure may cause a significant perturbation in the electronic structure of TiO_2 and, thus, enable the absorption of visible light. As shown in figure 15, cross-sectional TEM observations revealed that Vis-TiO_2 consists of large columnar crystals growing perpendicular to the substrate and the surface of the columnar crystals are covered with a stoichiometric TiO_2 phase. Such a stable surface phase for the columnar crystals may have worked as a passive phase, protecting the slightly reduced TiO_2 phases inside the bulk from complete oxidation [3-6,8-10].

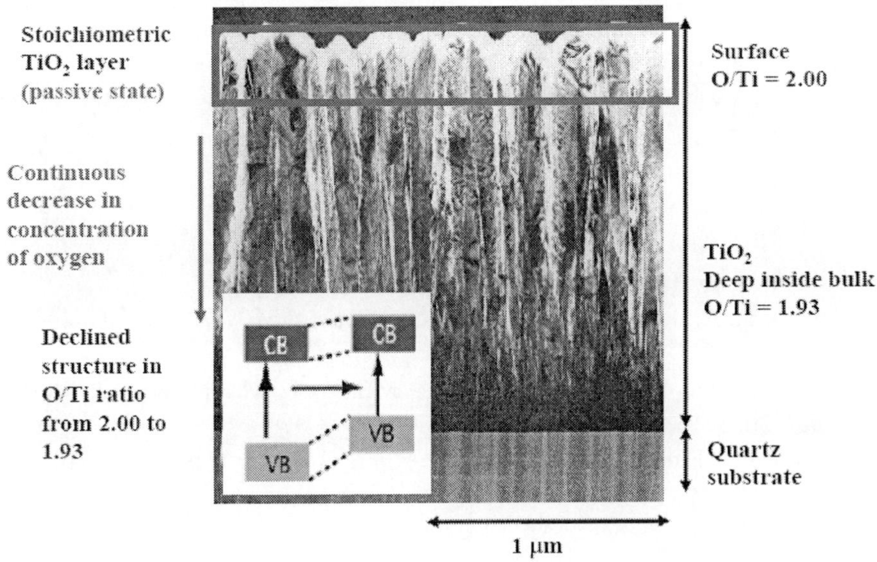

Figure 15. Cross-sectional TEM image of the Vis-TiO_2 thin film photocatalyst prepared on a quartz substrate.

5. Summary

The recent advances in the development of titanium oxide-based catalysts with high photocatalytic reactivity and photo-induced superhydrophilic properties were presented here for promising applications in solar energy conversion and environmental remediation. Special attention has been focused on the preparation of the highly dispersed titanium oxide species within zeolite cavities as well as the design of second-generation TiO_2 photocatalysts prepared by an advanced metal ion-implantation method which enables reactions even under visible light. A new and more cost-efficient RF magnetron sputtering deposition method for the preparation of visible light-responsive TiO_2 thin film photocatalysts was also presented along with detailed characterization studies and the various significant reactions that could be initiated. It was clearly demonstrated that a combination of conventional chemical preparation methods and advanced physical ion-engineering techniques can provide new approaches to the utilization of solar light as the most abundant and safe energy resource, significantly to address urgent environmental issues on a large and global scale. Such unique photocatalysts will also have great impact on solar energy conversion processes such as H_2 production from water or CO_2 reduction using artificial photosynthetic techniques.

References

[1] Fujishima, A.; Honda, K. *Nature.* 1972, 238, 37-38.
[2] Fujishima, A.; Rao, T. N.; Tryk, D. A. *J. Photochem. Photobiol.* C. 2000, 1, 1-21.
[3] Matsuoka, M.; Kitano, M.; Takeuchi, M.; Anpo, M.; Thomas, J. M. *Top. Catal.* 2005, 35, 305-310.
[4] Kitano, M.; Tsujimaru, K.; Anpo, M. *Appl. Catal. A: Gen.* 2006, 314, 179-183.
[5] Matsuoka, M.; Kitano, M.; Takeuchi, M.; Tsujimaru, K.; Anpo, M.; Thomas, J. M. *Catal. Today*. 2007, 112, 51-61.
[6] Kitano, M.; Matsuoka, M.; Ueshima, M.; Anpo, M. *Appl. Catal. A: Gen.* 2007, 325, 1-14.
[7] Anpo, M.; Che, M. *Adv. Catal.* 2000, 44, 119-257.
[8] Anpo, M.; Takeuchi, M. *J. Catal.* 2003, 216, 505-516.
[9] Anpo, M. *Bull. Chem. Soc. Jpn.* 2004, 77, 1427-1442.

[10] Anpo, M.; Dohshi, S.; Kitano, M.; Hu, Y.; Takeuchi, M.; Matsuoka, M. *Annu. Rev. Mater. Res.* 2005, 35, 1-27.
[11] Anpo, M.; Thomas, J. M. *Chem. Commun.* 2006, 31, 3273-3278.
[12] Anpo, M.; Matsuoka, M. Turning Points in Solid-State, Materials and Surface Science; RSC publishing: Cambridge, 2008; pp 492-506.
[13] Photofunctional Zeolites; Anpo, M.; Ed.; NOVA Science Publishers: New York, 2000.
[14] Matsuoka, M.; Anpo, M. Photochem. Photobiol. C: Photochem. Rev. 2003, 3, 225-252.
[15] Yamashita, H.; Ikeue, K.; Takewaki, T.; Anpo, M. *Top. Catal.* 2002, 18, 95-100.
[16] Anpo, M. *Pure Apple. Chem.* 2000, 72, (2000) 1265-1270.
[17] Takeuchi, M.; Anpo, M.; Hirao, T.; Itoh, N.; Iwamoto, N. *Surf. Sci. Jpn.* 2001, 22, 561-567.
[18] Yamashita, H.; Ichihashi, Y.; Anpo, M.; Hashimoto, M; Louis, C.; Che, M. *J. Phys. Chem.* 1996, 100, 16041-16044.
[19] Yamashita, H.; Harada, M.; Misaka, J.; Takeuchi, M.; Neppolian, B.; Anpo, M. *Catal. Today*. 2003, 84, 191-196.
[20] Dohshi, S.; Takeuchi, M.; Anpo, M. *Catal. Today*. 2003, 85, 199-206.

In: Environmentally Harmonious Chemistry… ISBN: 978-1-60876-428-0
Editors: M. Anpo and K. Mizuno

Chapter 4

ECO-FRIENDLY ELECTROCHEMICAL DEVICES USING "WATER" AND "HYDROGEN" AS KEY SUBSTANCES IN THE 21ST CENTURY

Hiroshi Inoue

ABSTRACT

"Water" and "hydrogen" are symbolic words of ecology and clean energy, respectively, and are key substances in the 21st century. Eco-friendly electrochemical devices incorporating either or both substances may be the most desirable power sources in this century. In this chapter, two eco-friendly electrochemical devices, the nickel/metal hydride battery and fuel cell, will be introduced. The former has already been commercialized and new applications are also being developed. On the other hand, the latter is still in the stage of fundamental research for practical use. The principles, features and outlook for the future of such eco-friendly electrochemical devices will be presented in this chapter.

INTRODUCTION

What is the most important substance for our life? You may have various answers to this question. I can say with certainly that it is "water". I believe nobody will deny this answer. There is plenty of water in nature. Sea water, rainwater, and wastewater which can be purified to clean and pure water by filtration and distillation. Water has various unique properties. For instance, water has an extraordinary high dipole moment and polarizability compared to organic solvents such as methanol, acetonitrile, dimethyl formamide, etc., leading to the high solubility of electrolytes and the formation of hydrogen bonds followed by high ionic conductivity. Water dissociates to hydrogen ions and hydroxide ions and its ionic product is 10^{-14}, which is the origin of acidity and basicity.

If voltage over 1.6 V is applied between two Pt electrodes in a sulfuric acid aqueous solution, hydrogen and oxygen evolve at the cathode and anode, respectively.

(Cathode) $2H^+ + 2e^- \text{-->} H_2(g)$ (1)

(Anode) $H_2O(l) \text{-->} 1/2O_2(g) + 2H^+ + 2e^-$ (2)

(Total) $H_2O(l) \text{-->} H_2(g) + 1/2O_2(g)$ (3)

Hydrogen is the smallest and lightest of all molecules on earth. Hydrogen burns to form water (or water vapor) as represented by a reverse reaction of Eq. (3), while organic compounds such as methanol and methane burn to form carbon dioxide, one of the greenhouse gases, together with water. The change in enthalpy ($\Delta H_{comb,H2}$) for the combustion of hydrogen (Eq. (4)) is -286 kJ mol^{-1}.

$H_2(g) + 1/2O_2(g) \text{-->} H_2O(l)$ (4)

The $\Delta H_{comb,H2}$ also means the change in enthalpy for the formation of water ($\Delta H_{f,H2O}$). In general, the change in Gibbs free energy (ΔG) is evaluated by subtracting the entropy term ($T\Delta S$) from the change in enthalpy (ΔH), that is $\Delta G = \Delta H - T\Delta S$. The change in Gibbs free energy for the formation of water ($\Delta G_{f,H2O}$) is -237 kJ mol^{-1} and it is consumed as heat energy during the combustion of hydrogen. The $\Delta G_{f,H2O}$ can be transformed into not only heat energy but also other energies. For example, devices transforming chemical energy, as expressed in Eq. (4), into electric energy are called “fuel cells”.

The boiling point of hydrogen is so low (ca. 20 K) that it usually becomes gas at room temperature. The density of hydrogen gas is also low (0.090 g dm^{-3} at 1 atm and 298 K). If we want to use hydrogen gas as fuel for a vehicle, we need a compact reservoir which can store sufficient amounts of hydrogen like a gasoline tank. The volume of hydrogen gas of 1000 m^3 under a pressure of 1 atm is 6.7 m^3 under a pressure of 150 atm. When the hydrogen gas is liquefied, the volume contracts to 1.3 m^3. Can we further compress the volume? A hydrogen storage alloy can realize such a compression. The hydrogen storage alloy stores a great amount of hydrogen in its crystal lattices to form hydride. For example, one mol of $LaNi_5$ alloy can store 6 mol of hydrogen. If the alloy stores hydrogen gas of 1000 m^3 under a pressure of 1 atm, the volume of the formed hydride is 0.9 m^3 which is around one thousandth of that!

When the hydrogen storage alloy is cooled in a hydrogen atmosphere and pressed at more than its equilibrium hydrogen pressure, it absorbs hydrogen to form hydride. Inversely, the heating or decompressing of the hydride causes the desorption of hydrogen from the hydride to revert back to the hydrogen storage alloy. Moreover, hydrogen absorption and desorption occur reversibly. The hydrogen storage alloy also has other unique properties. For example, the hydrogen storage alloy evolves heat during hydrogen absorption while the formed hydride desorbs heat during hydrogen desorption, which is an example of mutual transformation between chemical and thermal energies and applied for heat pumps and refrigerators, etc. The change in hydrogen pressure with hydrogen absorption and desorption can be transformed into mechanical energy which is an example of mutual transformation between chemical and mechanical energies and applied for compressors, sensors, and actuators, etc. Moreover, if the hydrogen storage alloy is used as an electrode, a mutual transformation device between the chemical and electric energies can be fabricated. That is realized as a nickel/metal hydride (Ni/MH) battery. In this way, we can say that the hydrogen storage alloy is a key material in mutual energy transformation.

In this chapter, two representative eco-friendly electrochemical devices, the Ni/MH battery and fuel cell, which involve “water” and “hydrogen” will be introduced.

Nickel/Metal Hydride Battery

1. Battery Reaction

The Ni/MH battery is composed of the hydrogen storage alloy/hydride (M/MH) negative electrode and the nickel oxyhydroxide/nickel hydroxide ($NiOOH/Ni(OH)_2$) positive electrode in an alkaline aqueous solution, as illustrated in figure 1. Both electrode reactions and battery reaction are represented as follows.

(Positive electrode) $Ni(OH)_2 + OH^- = NiOOH + H_2O + e^-$ (5)

$$\text{(Negative electrode)}\quad M + H_2O + e^- = MH + OH^- \qquad (6)$$

$$\text{(Battery reaction)}\quad Ni(OH)_2 + M = NiOOH + MH \qquad (7)$$

During charging, at the positive electrode $Ni(OH)_2$ is oxidized to NiOOH and, at the negative electrode, atomic hydrogen formed by water electrolysis is absorbed in M to form MH. On the other hand, during discharging at the positive electrode, NiOOH is reduced to $Ni(OH)_2$ and, at the negative electrode, MH is oxidized to M. The battery reaction is represented just by the transfer of hydrogen between both electrodes.

In a commercially sealed Ni/MH battery, the M/MH negative electrode has a higher capacity than the NiOOH/$Ni(OH)_2$ positive electrode or the capacity of the battery is controlled by that of the positive electrode. So during overcharging and overdischarging, gases like oxygen and hydrogen commonly evolve only at the positive electrode and are consumed at the negative electrode, as is explained below.

If the battery is overcharged, MH is still formed at the negative electrode along Eq. (6) while oxygen gas begins to evolve at the positive electrode (Eq. (8)). The oxygen gas diffuses through a separator[1] to the negative electrode in the electrolyte solution and chemically reacts MH to form water (Eq. (9)), leading to no rise of pressure in the cell.

$$\text{(Positive electrode)}\quad 2OH^- = H_2O + 1/2O_2 + 2e^- \qquad (8)$$

$$\text{(Negative electrode)}\quad 2MH + 1/2O_2 = 2M + H_2O \qquad (9)$$

If the battery is overdischarged, hydrogen gas begins to evolve at the positive electrode (Eq. (10)). The hydrogen gas diffuses through a separator to the negative electrode in the electrolyte solution and dissociates on M to atomic hydrogen (Eq. (11)), followed by a charge transfer reaction (Eq. (12)).

$$\text{(Positive electrode)}\quad 2H_2O + 2e^- = H_2 + 2OH^- \qquad (10)$$

$$\text{(Negative electrode)}\quad H_2 + 2M = 2MH \qquad (11)$$

$$2OH^- + 2MH = 2M + 2H_2O + 2e^- \qquad (12)$$

---

[1] The separator prevents the short-circuit of both electrodes without hindering the electrolyte ions and gases produced by overcharging and overdischarging from passing through it.

The Ni/MH battery not only has an overcharge protection mechanism but also an overdischarge protection mechanism. The Ni/Cd battery also has a similar overcharge protection mechanism to the Ni/MH battery. But the overdischarge protection mechanism is characteristic of the Ni/MH battery, which is a great advantage for high voltage batteries composed of some hundreds of cells for use in electric vehicles.

2. General Features

Since 1980, the progress of electronics has led to a rapid popularization of light and compact cordless appliances such as video cameras, mobile phones, laptop computers, shavers, toys, and tools, etc. [1], which has created a strong demand for compact, high power, long life and maintenance-free rechargeable batteries. At that time, the sealed Ni/Cd battery was commonly used, however further improvement of the energy density was an urgent problem, leading to the appearance of the Ni/MH battery. The sealed Ni/MH battery was first commercialized in 1990 by virtue of the development of new metal hydrides as a negative electrode material. An AA size Ni/MH cell has almost twice the capacity of a conventional Ni/Cd cell, leading to a long discharge period. Other advantages of the Ni/MH battery is summarized as follows:

1. An average discharge voltage of the Ni/MH battery of about 1.2 V, which is comparable to that of the Ni/Cd battery and a manganese primary battery.
2. High energy density
3. Self overcharge and overdischarge protection
4. Good high-rate chargeability and dischargeability
5. Eco-friendly active materials without Cd and Pb
6. High safety

Recently, by making use of these advantages, the Ni/MH battery has been applied for pure and hybrid electric vehicles and electric tools which require large discharge currents.

3. Hydrogen Storage Alloys- Negative Electrode Active Materials

Negative electrode active materials of the Ni/MH battery are metal hydrides. AB_5-type rare earth-based hydrogen storage alloys such as $Mm(Ni_{3.55}Mn_{0.4}Al_{0.3}Co_{0.75})_x$ (x = 0.9-1.0, Mm: misch metals) were first used for the commercial Ni/MH battery. Multi-component alloys have been developed considering their high capacity, light weight, high activity, long cycle life, and low cost, etc. $LaNi_5$ as an original material was known to have a high theoretical capacity of 372 mAh g^{-1}. However, it had a high cost, short cycle life and high equilibrium hydrogen pressure of over 1 bar at room temperature. La was partially replaced by misch metal, a mixture of rare earth metals like La, Ce, Pr, Nd, etc. due to its lower cost, however, the equilibrium hydrogen pressure increased. In order to lower it, Ni was partially replaced by Mn and Al which steadily dissolved during charge-discharge cycles. Moreover, Ni was also partially replaced by Co. Co played a role not in the decrease in the equilibrium hydrogen pressure but in the suppression of the dissolution of Mn and increased inner pressure and increase in cycle life. The optimization of the alloy composition led to $MmNi_{3.55}Mn_{0.4}Al_{0.3}Co_{0.75}$. The surface modification of the alloy, for example, the immersion of the alloy in alkaline solutions at high temperatures, led to longer cycle life. The negative deviation of B-site elements from 5 led to a decrease in the hydrogen equilibrium pressure, increase in discharge capacity and longer cycle life.

Co is more expensive than any other component in the alloy so that Co-less or Co-free alloys have been developed. Many Co-less or Co-free alloys have been proposed, but most of the alloys, except for $LaNi_{4.4}Cu$ [2] and $LaNi_{4.0}Al_{0.3}Si_{0.1}Fe_{0.6}$ [3], sacrifice the discharge capacity, cycle life and/or rate capability. Recently, Sanyo Electric Co. Ltd. (Japan) has developed a new hydrogen storage alloy and loaded it as a negative electrode active material in the latest Ni/MH batteries with 2500-2700 mAh capacity [4]. The alloy has a super-lattice structure, a regular stack of $CaCu_5$ and $MgNi_2$ structures, and a Ni/MH battery with the "super-lattice" alloy superior to the previous one in capacity, discharge at high currents (e.g., 3C[2]) and low temperatures (0 °C), charge at high temperatures (40 °C), and storage performance at high temperatures (40 °C).

[2] The "C" or "C rate" is a technical term in the field of battery technology and "nC" represents the current required for discharging a theoretical capacity of an alloy out for 1/n hours.

4. Outlook for the Future

The Nickel/MH battery has been replaced by the lithium ion secondary battery as a power source for portable electronic appliances such as video cameras, mobile phones, and lap-top computers, etc., since the second half of the 1990's since the latter has three times higher voltage per cell and much lighter weight than the conventional battery. However, lap-top computers with the lithium ion secondary battery have overheated and burst into flames since last year [5], and this has been responsible for delaying application to high-power uses such as HEVs. The safety of the Ni/MH battery is, therefore, being revised and improved, but the speculative jump of raw material cost has become a current issue.

Now, we should note two current applications for the Ni/MH battery. One is to use it like a dry cell battery. This new idea has been proposed by Sanyo Electric Co., Ltd. and the new Ni/MH battery is called "eneloop®" [6]. The appearance of the "eneloop®" is ascribed to the significantly improved self-discharge performance of the Ni/MH battery, and we can use it immediately after purchase as well as repeatedly without disposing of it. Another is to use it as on-site large-scale power sources for electric trains, buildings and the storage of electricity made by natural forces such as solar and wind. For this purpose, Kawasaki Heavy Industry Co., Ltd. Has developed a large-scale Ni/MH battery named "Giga cell®" and still attempting to improve battery performance.

FUEL CELLS

1. Battery Reaction

Reactions at both electrodes in fuel cells are inverse to the electrolysis of water represented by Eqs. (1)-(3).

$$\text{(Cathode)} \quad 1/2O_2(g) + 2H^+ + 2e^- \dashrightarrow H_2O(l) \qquad (13)$$

$$\text{(Anode)} \quad H_2(g) \dashrightarrow 2H^+ + 2e^- \qquad (14)$$

$$\text{(Total)} \quad H_2(g) + 1/2O_2(g) \dashrightarrow H_2O(l) \qquad (15)$$

Electrons move from the anode to the cathode through an external circuit, while current flows inversely. Such generation of electricity with the fuel cell using H_2

and O_2 was first demonstrated by Lord Grove in 1839. However, at that time, steam engines had been developed so that the fuel cell never attracted much attention for over 100 years. The standard electromotive force for a fuel cell is 1.23 V and its operating voltage is actually ca. 0.6-0.8 V, which is half of that of a dry cell battery due to the voltage drop called polarization. So fuel cells are usually stacked in series.

The theoretical energy conversion efficiency for a fuel cell using H_2 as a fuel is evaluated to be 83 % from the following equation (16):

$$\text{Theoretical energy conversion efficiency (\%)} = 100\Delta G/\Delta H \qquad (16)$$

where ΔG and ΔH represent the free energy and enthalpy changes for Eq. (15). In general, the efficiency for thermal power generation has difficulty exceeding 60 %, suggesting that the fuel cell is an energy saving power generation system. This is the reason we expect an increase in the practical use of the fuel cell.

2. General Features

The fuel cell is an apparatus converting chemical energy to electric energy like thermal power generation, however, the conversion process for the former is quite simple compared to the latter. In thermal power generation, the chemical energy is first converted to thermal energy by a boiler and then to mechanical energy by a turbine before generating the electric energy, leading to a lower efficiency for the generation of electricity. In contrast, in the fuel cell, the chemical energy is directly converted to electric energy, leading to lower energy loss and higher efficiency for the generation of electricity. While the fuel cell operates, not only electric energy but also thermal energy are generated because the entropy change of Eq. (15) is emitted as thermal energy. Therefore, the simultaneous use of both energies, which is called cogeneration, is recommended to raise the efficiency. Fuel cells for domestic use are designed to use thermal energy mainly for hot-water-supply and floor heating.

The general features of fuel cells are as follows:

1. They generate electricity as long as fuel and O_2 are supplied.
2. They hardly emit any environmental pollutants like NO_x, SO_x, CO_2, etc.
3. They have high efficiency for power generation, which does not depend on their scale and load factor.
4. They can easily be used to construct a cogeneration system.

5. They are quiet.
6. They are tolerant for a variety of fuels like natural gas, methanol, naphtha, etc.
7. They are so compact that we can use them on site.

3. Classification of Fuel Cells

Fuel cells are often classified in terms of the fuel, electrolyte and operating temperature. By classification in terms of the electrolyte, the following fuel cells are known:

1. Alkaline electrolyte fuel cell (AFC)
2. Phosphoric acid electrolyte fuel cell (PAFC)
3. Molten carbonate electrolyte fuel cell (MCFC)
4. Solid oxide electrolyte fuel cell (SOFC)
5. Solid polymer electrolyte fuel cell (PEFC) or Proton exchange membrane fuel cell (PEMFC)

An outline of each fuel cell is summarized in table 1.

Table 1. Outline of Various Fuel Cells

Type	Electrolyte	Fuel	Oxidant	Temperature (°C)	Efficiency for Power Generation (%)	Power Scale (kW)	Battery Reactions	Applications
AFC	Alkaline aqueous solution (OH^-)	Pure H_2	O_2 or air	RT – 230	60	1 – 250	(-) $H_2 + 2OH^- \rightarrow 2H_2O + 2e^-$ (+) $1/2O_2 + H_2O + 2e^- \rightarrow 2OH^-$ (Total) $H_2 + 1/2O_2 \rightarrow H_2O$	Space usages Military usages
PAFC	Phosphoric acid (H^+)	Reformed H_2	O_2 or air	180 - 205	35 – 45	50 - 10^4	(-) $H_2 \rightarrow 2H^+ + 2e^-$ (+) $1/2O_2 + 2H^+ + 2e^- \rightarrow H_2O$ (Total) $H_2 + 1/2O_2 \rightarrow H_2O$	On-site power generation
MCFC	Molten alkaline carbonate salts (CO_3^{2-})	Reformed H_2	(O_2 or air) and CO_2	630 - 670	45 – 60	10^3 - 10^5	(-)$H_2 + CO_3^{2-} \rightarrow CO_2 + H_2O + 2e^-$ (+) $1/2O_2 + CO_2 + 2e^- \rightarrow CO_3^{2-}$ (Total) $H_2 + 1/2O_2 \rightarrow H_2O$	Large scale power generation
SOFC	Stabilized zirconia (O^{2-})	Reformed H_2	O_2 or air	800 - 1000	50 – 60	10^3 - 10^5	(-) $H_2 + O^{2-} \rightarrow H_2O + 2e^-$ (+) $1/2O_2 + 2e^- \rightarrow O^{2-}$ (Total) $H_2 + 1/2O_2 \rightarrow H_2O$	Large scale power generation
PEFC	Proton exchange membrane (H^+)	Reformed H_2	O_2 or air	RT - 100	35 – 45	50 - 10^4	(-) $H_2 + 2OH^- \rightarrow 2H_2O + 2e^-$ (+) $1/2O_2 + H_2O + 2e^- \rightarrow 2OH^-$ (Total) $H_2 + 1/2O_2 \rightarrow H_2O$	Domestic use Vehicles On-site power generation

4. PEFC

PEFC is expected to be used for fuel cell vehicles (FCVs) and domestic uses. In a PEFC, a proton exchange membrane (Nafion®) having high chemical stability is used as an electrolyte and gas diffusion electrodes with a three-layer structure are used for both electrodes. The gas diffusion electrodes are prepared by three steps, *i.e.*, hydrophobic (wet-proofing) treatment of a carbon paper (CP), coating of a gas diffusion layer (GDL), and coating of the catalyst layer (CL) onto the GDL. The GDL is formed by coating a slurry of carbon black and PTFE-dispersions on CP. The CL with a Nafion ionomer is formed over the GDL. The catalyst paste is prepared by mixing the catalyst powder, Nafion® solution and solvent, which is called a catalyst ink. The catalyst ink is uniformly spread over the GDL. Finally, the Nafion® membrane is sandwiched by the gas diffusion electrodes, followed by hot-pressing, which is called membrane/electrode assembly (MEA). The typical catalyst powder is prepared by loading Pt nanoparticles for the cathode or Pt/Ru alloy nanoparticles for the anode on the carbon black particles.

Development of the PEFC was started by General Electric Co. in the USA in the latter half of the 1950's. In the 1960's, the Nafion® membrane was invented by Du Pont Co., leading to the prolonged life of the PEFC. The PEFC was mainly used for space and military uses in the beginning of development. In 1980, Ballard Power System Co. in Canada started the development of PEFCs for general use. In the 1990's, research and development on PEFCs for FCVs and domestic uses increased dramatically since we faced serious environmental problems such as global warming and our consciousness of environmental preservation was significantly enhanced.

The features of the PEFCs are summarized as follows:

1. They have a compact size, light weight and high power.
2. The warm-up time is short and inexpensive components can be used due to operation at lower temperatures.
3. Maintenance is simple.
4. The pressure of both electrodes can be regulated separately.

However, there are a lot of technological problems for each component in the PEFC. For example, on the electrode catalyst, an improvement of the catalytic activity of the cathode catalysts and the CO-tolerance of the anode catalysts is desired, while on the electrolyte membrane, an improvement of ionic conductivity and chemical stability at higher temperatures would be significant.

To commercialize PEFCs, cost cuts for each component and improvement of their durability are urgent and important. In particular, the preparation of Pt-less and Pt-free catalysts and alternative electrolyte membranes to Nafion® are hot topics around the world. Moreover, fuel cells using H_2 and O_2 need a storage tank of high-pressure H_2. For use in FCVs, tanks which can safely store H_2 with a pressure of 35 MPa (70 MPa in the near future) have been produced and loaded.

5. Outlook for Future

When H_2 burns, CO_2 is not produced like fossil fuels and water is the sole product. However, we should remember that H_2 is produced by reforming hydrocarbons like methane and CO_2 is then produced. However, it is estimated that FCHs using H_2 produced from a natural gas reduces the amount of CO_2 exhausted by ca. 60 % compared to present gasoline vehicles. If H_2 is produced from nonfossil fuels, the amount of CO_2 exhausted would be zero. For example, water electrolysis using solar cells and the use of semiconductor photocatalysts and biomass, etc., are desirable H_2 generation methods but they are yet far from practical use. Nevertheless, the use of hydrogen energy, the key energy resource in the 21st century, is greatly useful in suppressing CO_2 exhaust and the role of the fuel cell is also significant.

REFERENCES

[1] Sakai T., Matsuoka M. & Iwakura C. (1995) Handbook on the Physics and Chemistry of Rare Earths Vol. 21: Rare Earth Intermetallics for Metal-Hydrogen Batteries. In: Gschneidner, Jr. K. A. & Eyring L. (Eds.), (pp. 133-178). Amsterdam, Elsevier Science B. V.

[2] Notten, P. H. L., Einerhand, R. E. F., Daams, J. L. C. (1994). On the nature of the electrochemical cycling stability of non-stoichiometric $LaNi_5$-based hydride-forming compounds Part I. Crystallography and electrochemistry, *J. Alloys Compd.,* 210, 221-232.

[3] Iwakura, C., Ohkawa, K., Senoh, H. & Inoue, H. (2002). A Co-free AB_5-type hydrogen storage alloy for nickel-metal hydride batteries: $LmNi_{4.0}Al_{0.3}Si_{0.1}Fe_{0.6}$, *J. Electrochem. Soc.,* 149, A462-A465.

[4] Yasuoka, S., Magari, Y., Murata, T., Tanaka, T., Ishida, J., Nakamura, H., Nohma, T., Kihara, M., Baba, Y., Teraoka, H. (2006). Development of

high-capacity nickel-metal hydride batteries using superlattice hydrogen-absorbing alloys, *J. Power Sources,* 156, 662-666.

[5] Chemical & Engineering News, (2006) Aug. 21, (p.11) American Chemical Society.

[6] http://www.sanyo.co.jp/koho/hypertext4-eng/0511/1101-2e.html

In: Environmentally Harmonious Chemistry… ISBN: 978-1-60876-428-0
Editors: M. Anpo and K. Mizuno

Chapter 5

ENVIRONMENTALLY HARMONIOUS ORGANIC PHOTOCHEMICAL REACTIONS

Kazuhiko Mizuno, Hiroshi Ikeda and Hajime Maeda

ABSTRACT

In view of the importance of environmentally benign chemistry, we have developed a variety of organic photochemical reactions. In this chapter, our recent advances in synthetically useful and mechanistically interesting photochemical reactions are described in detail. Section 5-2 deals with the highly stereoselective inter- and intra-molecular photocycloaddition of alkenes to aromatic rings via exciplex. In Section 5-3, synthetically useful photoreactions by use of microreactors are described. The first part of Section 5-4 shows photoinduced electron transfer reactions including carbon-carbon bond forming reactions using carbon-nucleophiles or organosilicon compounds. The second part of this section describes the photorearrangement and photooxygenation of arylated methylenecyclopropanes from mainly mechanistic viewpoints. In this section, thermoluminescence and organic radical light-emitting diode of these compounds are also described. Section 5-5 deals with highly emissive organosilicon compounds.

5.1. Introduction

Photochemical reactions are convenient and useful methods of environmentally harmonious organic synthesis. Synthesis using photoenergy has the following advantages: (1) It is easy to control the chemical reactions in selection of excitation wavelength, solvents, reaction temperature, and reaction phase. (2) Photoreaction occurs smoothly compared with thermal reaction, because reactive species are more oxidative or reductive than those of ground state molecules. (3) The photoinduced electron transfer (PET) process is a useful method for the generation of ionic species under neutral conditions. (4) Highly strained compounds can be sometimes prepared in one-step, and the key-step in total synthesis of natural products can often be utilized. (5) The reaction mechanism can be easily clarified by use of some techniques such as laser flash photolysis. From these points of view, we have developed a variety of synthetically useful and mechanistically interesting photoreactions. This chapter deals with our recent progress on photochemical reactions directed to green chemistry.

5.2. Photocycloaddition Reactions

The photocycloaddition reaction is not only one of the most useful methods for the preparation of cyclic compounds in a one-step process, but also for high atom-efficiency in view of green chemistry. In this subsection, we will discuss inter- and intramolecular photocycloadditions. The photocycloaddition of alkenes to aromatic rings such as benzene and naphthalene derivatives can be simply classified as [2+2], [3+2], and [4+2] types, depending on the substituents, reaction media, and additives (Scheme 1). Most photoreactions occur to give [2+2] cycloadducts at normal 1,2-position. However, in some cases, unusual cycloadducts at the 1,8a-, 1,8-, and 1,3-position are obtained in the photoreaction of naphthalene derivatives.[1]

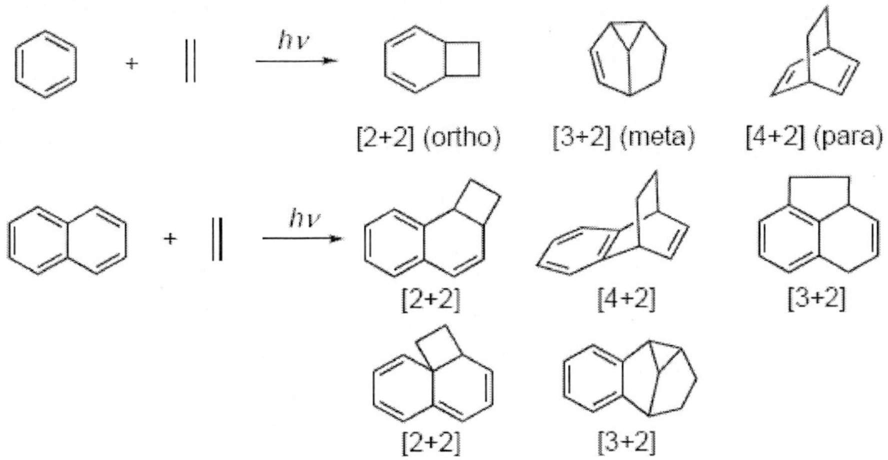

Scheme 1. Classification of the photocycloaddition of alkenes to benzene and naphthalene rings.

5.2.1. Stereoselective Inter- and Intramolecular Photocycloaddition

Irradiation of a benzene solution containing 1-cyanonaphthalene (1) and *trans*- and *cis*- phenyl propenyl ether (*trans*-2 and *cis*-2) through a Pyrex filter afforded [2+2] photocycloadducts 3 and 4 in a stereospecific and *endo*-selective manner (Scheme 2). However, *endo*-selectivity is not observed in the photoreaction of ethyl vinyl ether (6) to 1. Interestingly, the photoreaction of 1-cyano-4-hydroxymethylnaphthalene (5a) with 6 gave a [2+2] photocycloadduct 7 in a highly *endo*-selective manner. This *endo*-selectivity can be reasonably explained by hydrogen bonding. Similar irradiation of 5a with 3-methyl-2-butenol afforded an *endo*-selective [2+2] photocycloadduct 9. Lowering the reaction temperature to –60 °C increased the ratio of *endo*-selectivity (92:8). The contribution of the hydrogen bond to the *endo*-selectivity was supported by low-temperature ^{1}H NMR.[2a-c]

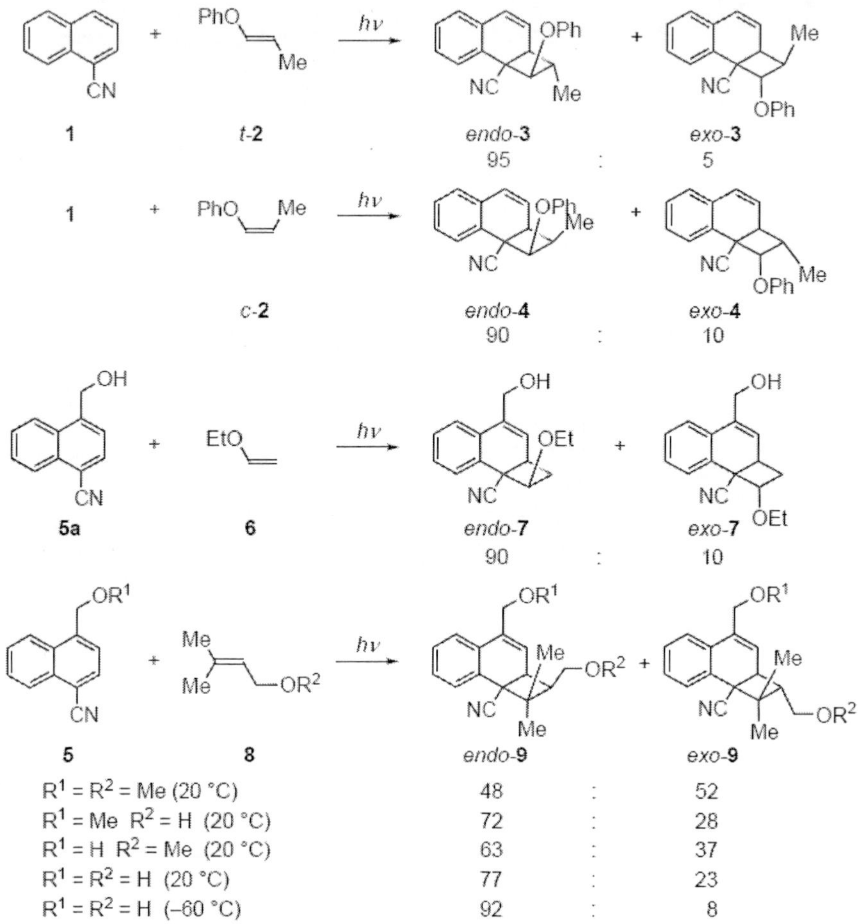

Scheme 2. Stereoselective photocycloaddition of alkenes to the naphthalene ring.

Photocycloaddition of 1 and 2-cyanonaphthalene (12) to furan has proceeded to give [4+4] and caged products, respectively. The different regioselectivity is elucidated by the difference in the dipole moment in the excited states between 1 and 12. We have found a specific example in the photocycloaddition of furan derivatives to cyanonaphthalenes. Irradiation of 1 with furfuryl methyl ether (10a) in benzene gave two isomeric [4+4] cycloadducts (*syn*-11 and *anti*-11) in a 4:1 ratio (Scheme 3). However, similar photoreaction of 1 with furfuryl alcohol (10b) afforded a sole isomer, *syn*-11. The fluorescence quenching rate of 1 by 10b in cyclohexane is five times faster than that by 10a. In addition, an emission of the exciplex of 1 and 10b is observed, being much stronger than that between 1 and 10a.

The photoreaction of 12 with 2-methylfuran gave two caged products in almost 1:1 ratio. Similar irradiation of 12 with 10 gave the corresponding caged products in 1:2 ratio. Interestingly, two isomeric products *syn*-14 and *anti*-14 were obtained in the photoreaction of 3-furylmethanol (13) to 12 in 5:1 ratio.[2d]

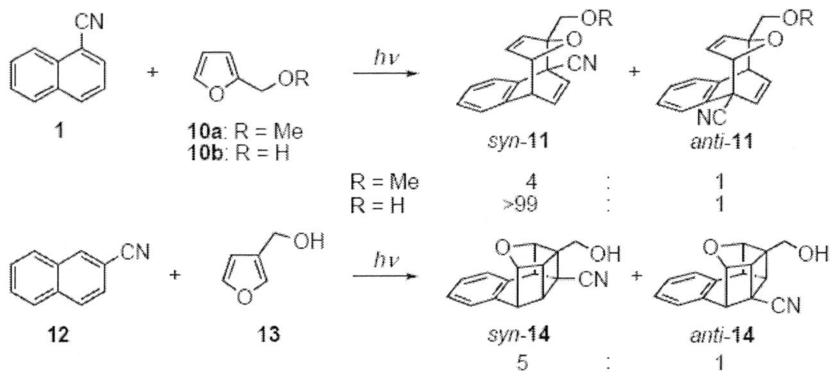

Scheme 3. Stereoselective photocycloaddition of furans to the naphthalene ring.

9-Cyanophenanthrene (15) reacts not only with electron-rich arylalkenes, but with electron-deficient arylalkenes. Irradiation of 15 with anethole 16 in benzene exclusively afforded a [2+2] cycloadduct 17 in a stereospecific and *endo*-selective manner (Scheme 4). In this photoreaction, a charge-transfer type exciplex is proposed by exciplex quenching experiments. The exciplex fluorescence is selectively quenched by dimethyl acetylene dicarboxylate, and almost the same quenching rate for the disappearance of [2+2] cycloadduct is observed. Photoreaction of cinnamonitrile (18a) and methyl cinnamate (18b) to 15 gave [2+2] cycloadducts 19a-b in a stereospecific manner. The π–π interaction between two electron-deficient aromatic rings is proposed in an exciplex intermediate.[3a-b]

Photoreaction of pyrene (20) with electron-deficient arylalkenes such as 18a and 18b afforded 1:1 cycloadduct 21 and 1:2 cycloadduct 22 in a stereospecific and *endo*-selective manner in high yields (Scheme 5). On prolonged irradiation, 1:2 cycloadduct 22 was precipitated.[3c]

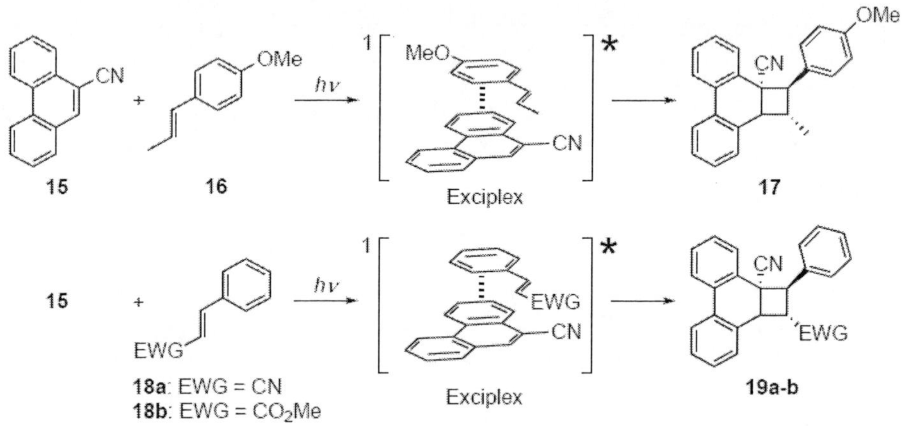

Scheme 4. Stereospecific photocycloaddition of arylalkenes to 9-cyanophenanthrene (15).

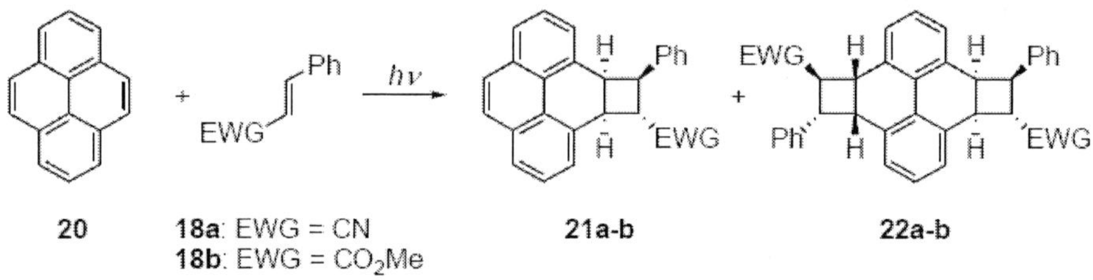

Scheme 5. Intermolecular photocycloaddition of arylalkenes to pyrenes.

5.2.2. Construction of Polycyclic Compounds via Intramolecular Photocycloaddition

Intramolecular photocycloaddition is not only a useful method for construction of polycyclic compounds, but also a methodology for the discovery of novel photoreactions. We have developed a variety of intramolecular photocycloadditions of bifunctional molecules with electron-donor and acceptor moiety. For example, irradiation of a benzene solution of 23a afforded an intramolecular cycloadduct 24a exclusively. Similar irradiation of 23b gave 24b in a siteselective manner. Conversely, 23c did not give any product (Scheme 6).[3d]

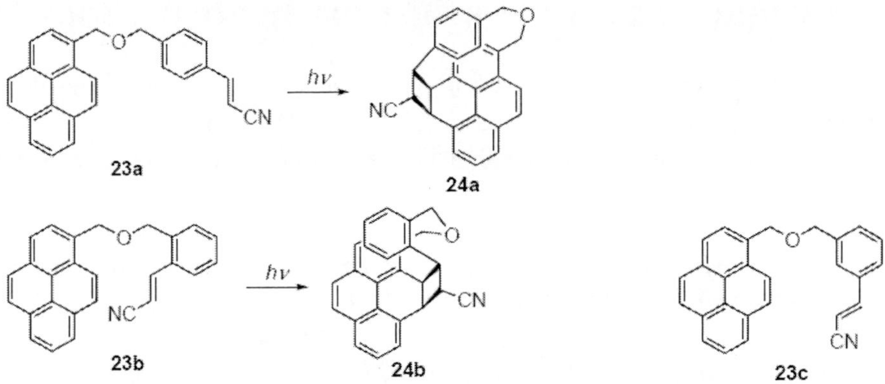

Scheme 6. Intramolecular photocycloaddition of arylalkenes to pyrenes.

Irradiation of a benzene solution containing 1-cyano-2-(5-methyl-2-oxa-4-hexenyl)naphthalene afforded a [2+2] photocycloadduct 25 at 1,2-position as a major product 26 (Scheme 7). However, this cycloadduct is smoothly returned to the starting material. Prolonged irradiation gave a [3+2] cycloadduct 27 exclusively. Similar intramolecular photocycloaddition of related compounds 28 afforded a variety of penta- and hexa-cyclic compounds 29 in good yields.[4]

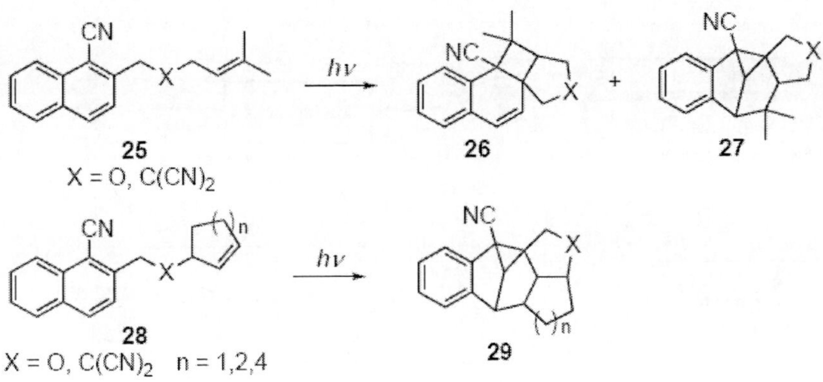

Scheme 7. One-step synthesis of benzotriquinanes and related compounds via the intramolecular photocycloaddition of alkenes to naphthalenes.

5.3. Photoreactions by Use of Microreactors

There are a variety of vessels for photoreactions. In general, preparative organic photoreactions are conducted in tubes or cells made of Pyrex or quartz, whose widths and depths are centimeter sizes. Recently, microreactors with narrow channel space whose width and depth are micrometer sizes have attracted much attention as new reaction vessels for not only photoreactions, but also thermal reactions.[5] There are two attractive features of the microreactors. The first one is the narrow depth. On the basis of Lambert-Beer's law ($A = \log (I_0/I) = \varepsilon cl$), in which A, I_0, I, ε, c, and l represent absorbance, intensity of incident light, intensity of transmitted light, molar extinction coefficient, concentration, and length of optical path, using values of $c = 1.0 \times 10^{-2}$ molL^{-1} and $\varepsilon = 10^4$ mol^{-1}Lcm^{-1}, 99.9% of incident light is absorbed to the solution from the surface of the vessel to the depth of 0.3 mm. If you use 10 mmϕ Pyrex tube for photoreactions, a greater part of solution cannot absorb the light under batch conditions. In other words, solutions in microreactors consume the energy of incident light just thoroughly. The other attractive feature is a flow system. In preparative photoreactions in which undesired secondary reaction competitively takes place, flow system is useful because it can eject the primary product from the reaction system with an appropriate flow rate.

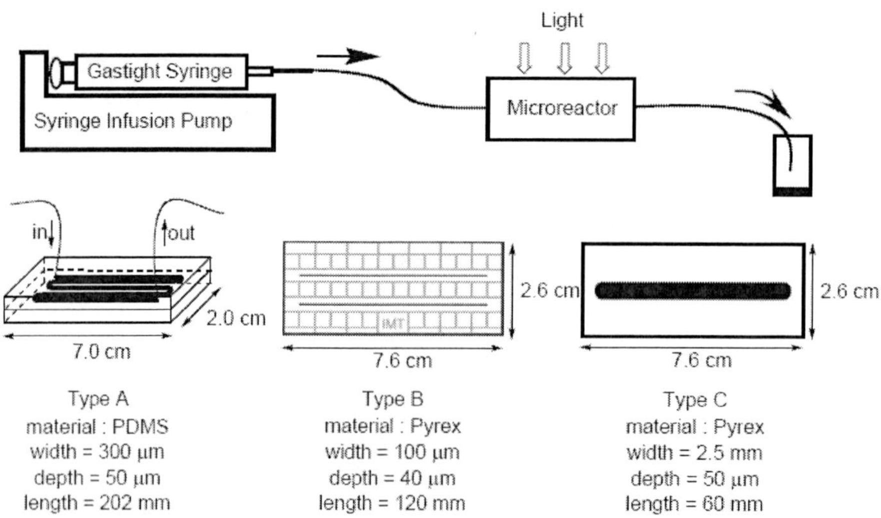

Figure 1. Apparatus for the photoreaction in the micro-flow system.

The requirement of the microreactors for photoreactions is transparency. From the viewpoint of easy construction and free design in one's own way, poly(dimethylsiloxane) (PDMS) and Pyrex glass are often chosen for materials for photochemical microreactors. Our microreactor apparatus with a flow-system for photoreactions is shown in figure 1. The Type A reactor made of PDMS can be easily prepared.[5e] Pyrex glass spin-coated by a positive type photoresist was irradiated by a tungsten lamp through a contacted photomask. The development and wet-etching by an HF-NH_4F aqueous solution gave a template having a protruded curve. The PDMS was molded by the template and was attached to another flat PDMS plate. Finally, capillary tubes (diameter = 100 μm) were inserted to furnish the Type A reactor. The Type B reactor is commercially available. The Type C reactor is made by hand by attaching an ionomer resin film with a bored hole in the center by two Pyrex glass plates.

We chose the intramolecular photocycloaddition of 1-cyano-2-oxaalkenylnaphthalene derivative 25a as a reference photoreaction (Scheme 8).[6] McCullough and our group have previously reported that the photoirradiation of benzene and acetonitrile solutions containing compound 25a afforded photocycloadducts at the 1,2-position (26a) and 2,4-position (27a) of the naphthalene ring.[7] Compound 26a was predominantly obtained at the initial stage of the reaction, but prolonged irradiation afforded 27a as a major product, because photocycloreversion from 26a to 25a takes place although that from 27a to 25a does not occur. This reversibility is supported by the UV absorption spectra of 25a-27a. Under the present irradiation conditions using PDMS with a UV-29 filter or Pyrex glass, 25a and 26a competitively absorb UV light but 27a does not absorb incident light.

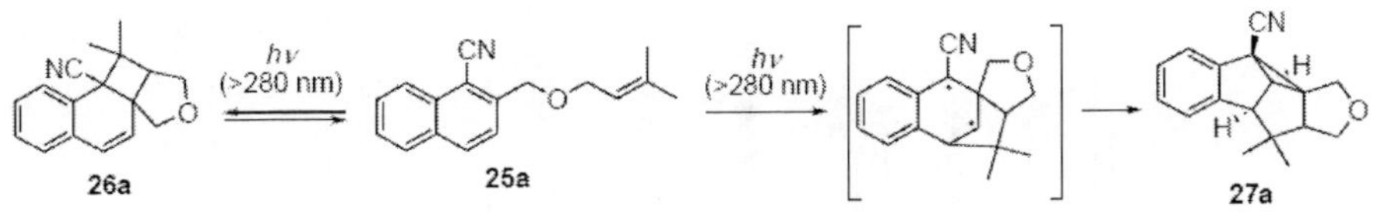

Scheme 8. Selective formation of [2+2] photocycloadduct using microreactors.

We found that, when the photoreaction was carried out in a microreactor under a flow system, both the regioselectivity of the products and the efficiency were higher than those under batch conditions.[6] The time dependency on the ratio of 25a and photoproducts 26a and 27a in the photoreaction of 25a in benzene solutions is shown in figure 2. Under batch conditions, after 1 h irradiation, half of 25a was consumed and 26a and 27a were formed in 46 and 8% yields, respectively (figure 2a). After 5-h irradiation, 25a decreased to 19% and the ratio of 26a and 27a reached about 3 : 2. On the other hand, in micro-flow system (Type B), when the residence time was 25 s, 25a decreased to less than the half and 26a and 27a formed in 60 and 2% yields, respectively (figure 2b). When the residence time of the solution was prolonged to 172 s by reducing the flow rate to 0.01 mL/h, the conversion of 25a became nearly equal to the conversion after 5-h irradiation under the batch conditions, but the ratio of 26a and 27a was about 12 : 1. These data suggest that a difficulty in obtaining the initial photoproduct exclusively on conventional batch system can be overcome by using the micro-flow system.

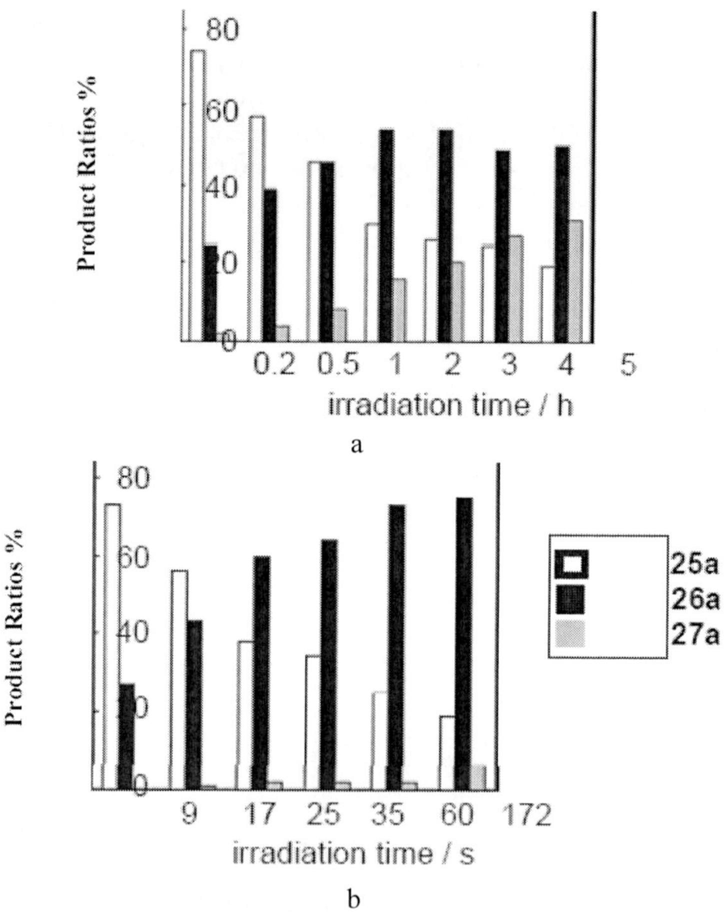

Figure 2. Product ratios upon photoirradiation of benzene solutions of 25a. (a) under batch conditions and (b) in micro-flow conditions (Type B).

Figure 3 shows a schematic representation of the effects of the microreactor-flow system as compared with the batch system. Depict in mind a case where the primary photoproduct, which is produced by photoirradiation of the substrate, reacts competitively upon photoirradiation to give undesired secondary product(s). Under the batch conditions without stirring, the primary product is produced only on the surface of the vessel at the initial stage of the photoreaction, then a secondary reaction proceeds faster than the diffusion of the primary product which has already accumulated near the surface. On the other hand, in the flow system, the selectivity of products can be improved by the quick outflow of the primary product from the reaction system. From the small Reynolds number of these microreactors, it is also anticipated that the laminar flow might be predominant in microreactors, so the mixing effect arising from a turbulent flow might be negligible. The increment of reaction efficiency can be attributed to the thorough absorption of the incident light. It is also notable that the flow rates can be liberally chosen by the reactivity of the substrates. Namely, a slower flow rate increased the conversion, and faster flow rate decreased the conversion. Thus, in such a reaction system in which both a fast reversible process and a slow irreversible process coexist, a micro-flow system is appropriate for the selective formation of the initial product.

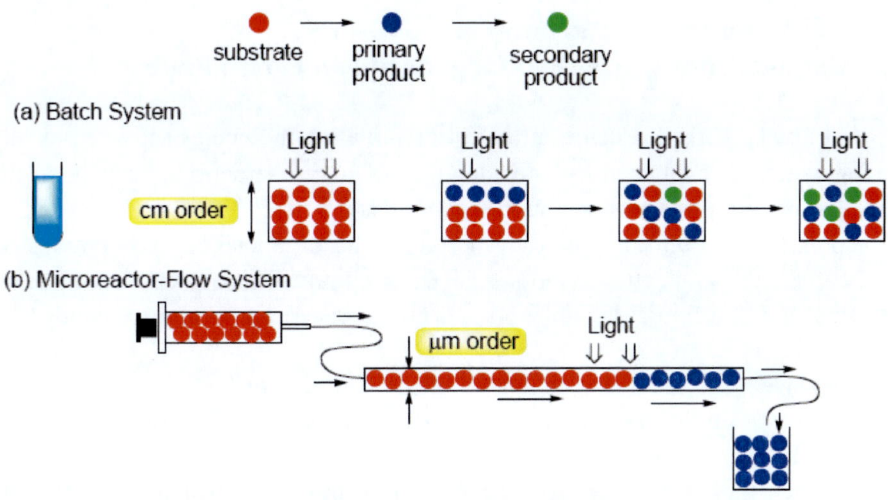

Figure 3. Effects of a microreactor-flow system compared with a conventional batch system.

As one of the weak points of microreactors, the quantity of products that could be obtained was too small. To overcome the problem, the Type C microreactor with a wider width (2.5 mm) than Type B (100 μm) was developed without a significant change in the depth. As a result, a large quantity of products was obtained with a similar regioselectivity when Type C reactor was used under the same rate and irradiation time as those in the Type B reactor.

5.4. Photoinduced Electron-Transfer Reactions

Photoinduced electron-transfer (PET) reactions between electron donating molecules (D) and electron accepting molecules (A) serve as useful methods for the generation of radical ion species under neutral and mild conditions.[8] In less polar solvents such as benzene, exciplex intermediates are produced, as described in section 5-2. In polar solvents such as CH_3CN, efficient electron transfer occurs from D to A* (or from D* to A) to give $D^{\bullet+}$ and $A^{\bullet-}$. These ionic species produce a variety of products and mechanistically interesting phenomena. In this section, some typical examples of PET reactions are described.

5.4.1.1. Photoinduced Carbon-Carbon Bond Forming Reactions by Use of Organosilicon Compounds

Allylic and benzylic silanes are electron donating molecules activated by silyl groups, which is explained by consideration of σ–π interactions. In general, the oxidation potentials of organosilicon compounds are much lower than those of the corresponding carbon compounds. Therefore, the organosilicon compounds are good candidates as electron donors for PET reactions. This sub-section describes photoinduced carbon-carbon bond forming reactions using organosilicon compounds.[9]

We have reported the efficient photosubstitution of *p*-dicyanobenzene (*p*-DCB, 30) by allylic silanes in CH_3CN (Scheme 9).[10] The photoreaction is triggered by one-electron transfer from allyltrimethylsilane (31) to the excited singlet state of 30 to give the radical cation of 31 and the radical anion of 30 ($30^{\bullet-}$). The second step is the formation of allyl radical and silyl cationic species. The next step is an *ipso*-attack of allyl radical to $30^{\bullet-}$ to give an anionic intermediate. A decyanation then gives 1-allyl-4-cyanobenzene (32) with a high yield. It is notable here that the C-Si bond cleavage is assisted by the attack of a nucleophile such as

CH_3CN on the silicon atom. This conclusion is supported by the observation that the quantum yield for the formation of 32 decreases with the bulkiness of the silyl group in the order $SiMe_3 > SiEt_3 > SiMe_2{}^tBu > Si^iPr_3 > SiPh_3$.

A similar photoreaction of 1,4-dicyanonaphthalene (33) with 31 primarily afforded a reductive allylation product 34, which is followed by photocycloaddition to give a caged product 35 in a high yield.

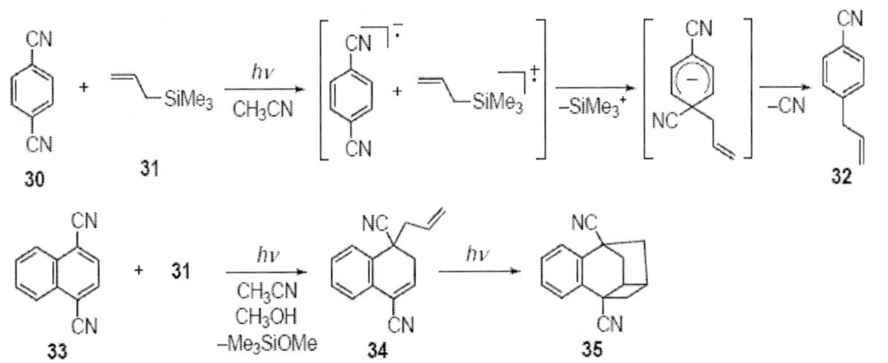

Scheme 9. Regioselective photosubstitution and photoaddition using allyltrimethylsilane (31).

We have also found that the photoreaction of 1,1-dicyano-2-phenylethene (36) with allylsilane 31 in the presence of phenanthrene in CH_3CN affords 5,5-dicyano-4-phenyl-1-pentene (37) with a good yield (Scheme 10). Phenanthrene acts as a redox photosensitizer and allylation occurs at the benzylic position. On the other hand, similar photoallylation of cyclohexylidenepropanedinitrile (38) occurs at the position α to the cyano group in the presence of acetic acid or formic acid. In the absence of these acids, dehydroallylation product 40 and reduction product 41 were obtained as main products. Therefore, this photoreaction proceeds in a pK_a-dependent manner.[10e]

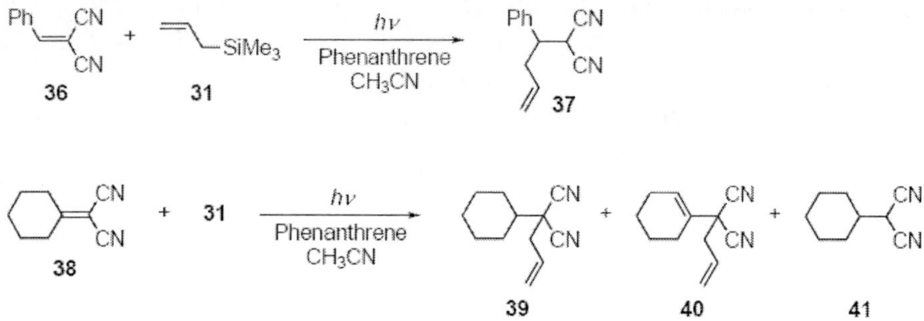

Additive	pK_a	Ratio 39	40	41
Molecular sieve 4A		0	44	56
CH_3OH	15.0	4	29	67
$CH_2(CO_2Me)_2$	13.5	19	17	64
CH_3CO_2H	4.8	69	0	31
HCO_2H	8.0	92	0	8

Scheme 10. Regioselective photoallylation using allyltrimethylsilane.

5.4.1.2.Photoinduced Carbon-Carbon Bond Forming Reactions by Use of Carbon Nucleophiles

PET reactions of alkenes in the presence of nucleophiles have been widely investigated in these recent four decades. However, little is known about the attack of carbon nucleophiles to radical cation species. Very recently, we have found that irradiation of an aqueous CH_3CN solution containing propanedinitrile, 1,1-diphenylethene (42), lithium carbonate, and a catalytic amount of 9-cyanophenanthrene (15) leads to *anti*-Markovnikov addition to form monoalkylated propanedinitrile 43 in a high yield (Scheme 11).[11] A similar photoreaction of 2,5-dimethyl-2,4-hexadiene (44) afforded the dimeric product 45 with a moderate yield. Since the selective carbon-carbon bond forming reaction is an important methodology in organic synthesis, this photoreaction has the potential of being an environmentally benign method.

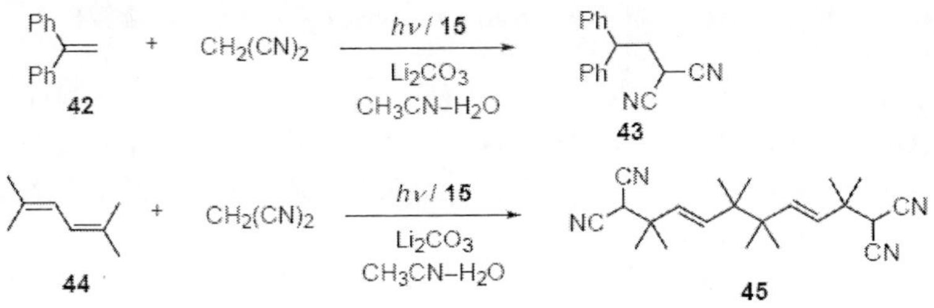

Scheme 11. Carbon-carbon bond formation via the PET process.

5.4.2. Rearrangement and Oxygenation of Arylated Methylenecyclopropane (MCP) Triggered by Photoinduced Electron Transfer (PET)

Heat treatment of the parent methylenecyclopropane (MCP) generates a trimethylenemethane (TMM),[12] a representative of non-Kekulé molecules, which quantitatively regenerates the starting MCP. This reaction is called a degenerate MCP rearrangement and has been observed for the 2,2-diaryl derivatives (46a,b, Scheme 12) in a low yield under direct photoirradiation conditions[13] and in a high yield under PET conditions using an electron-accepting photosensitizer (sens).[14,15] In this section, we have focused on the degenerate MCP rearrangement and oxygenation of 46 triggered by PET with a variety of sens.

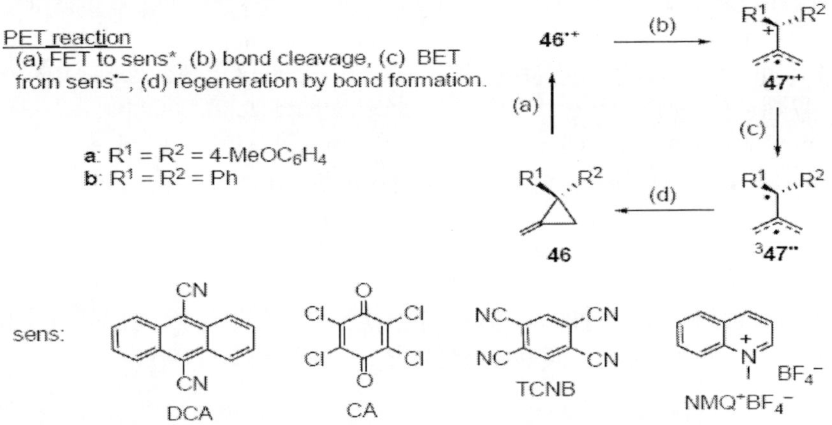

Scheme 12. A mechanism of PET degenerate MCP rearrangement of 46.

5.4.2.1. PET Degenerate MCP Rearrangement of Arylated MCP

The name of the "degenerate" MCP rearrangement comes from the fact that unlabeled 46 does not give any product(s). The reactivity of 46 under PET conditions using 9,10-dicyanoanthracene (DCA) as a sens is then confirmed with the labeled 46, d_2-46 (Scheme 13), which gave a photostationary mixture of d_2-46 and d_2-46' in ~50:50 ratio.[16] Previous study of chemically induced dynamic nuclear polarization (CIDNP) of 46 under the chloranil (CA)-sensitized PET conditions confirmed that a TMM-type radical cation ($47^{\bullet+}$) are generated as a key intermediate in the reaction.[14] Recently, however, we discovered that not only $47^{\bullet+}$ but also the corresponding TMM-type biradical ($47^{\bullet\bullet}$) also intervenes in the reaction.[15]

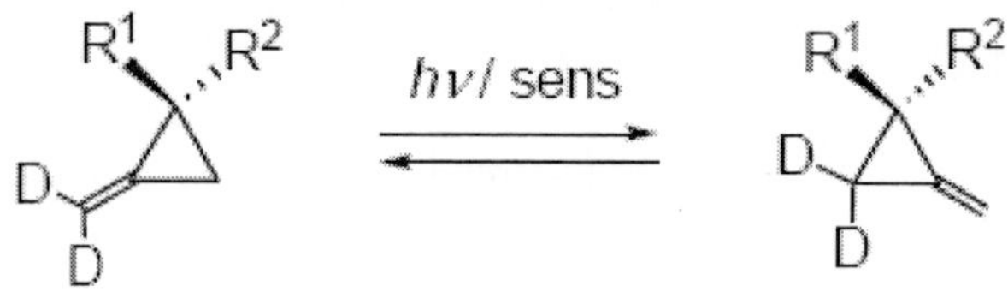

Scheme 13. PET degenerate MCP rearrangement of d_2-46.

A mechanism for PET reactions of 46 consists of four steps (Scheme 12) and involves (a) *forward* electron transfer (FET) from 46 to sens* to produce $46^{\bullet+}$ and sens$^{\bullet-}$, (b) bond cleavage of $46^{\bullet+}$ to give $47^{\bullet+}$, (c) *backward* electron transfer (BET) from sens$^{\bullet-}$ to $47^{\bullet+}$ yielding $^{3}47^{\bullet\bullet}$, and (d) quantitative regeneration of 46 from $^{3}47^{\bullet\bullet}$ by bond formation.[15] A key mechanistic step in this pathway is step (c), which has been fully characterized and summarized, as described below. In particular, two different types of TMM-type intermediates ($47^{\bullet+}$ and $^{3}47^{\bullet\bullet}$) were detected using time-resolved absorption and electron paramagnetic resonance (EPR) spectroscopy. Both $47^{\bullet+}$ and $^{3}47^{\bullet\bullet}$ exhibited the considerably twisted molecular geometries and localized electronic structures. Accordingly, the interaction between the bis(4-methoxyphenyl)methyl cation (or radical) subunit and the allyl radical subunit of $47^{\bullet+}$ (or $^{3}47^{\bullet\bullet}$) are quite small.[15,17]

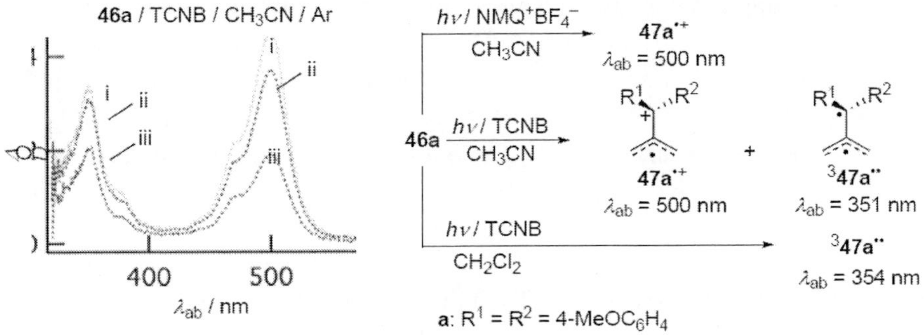

Figure 4. (Left) Time-resolved absorption spectra observed at a delay time of 200 ns (i), 1 μs (ii), and 5 μs (iii) after laser flash photolysis (308 nm excitation) of 46a under TCNB-sensitized PET conditions in CH_3CN. (Right) The results of laser flash photolyses of 46a that depends on the sens and solvent employed.

The time-resolved absorption spectra of 47a$^{•+}$ ($R^1 = R^2$ = 4-$MeOC_6H_4$, λ_{ab} = 500 nm in CH_3CN, figure 4) and ^{3}47a$^{••}$ [λ_{ab} = 351 nm in CH_3CN (figure 4) or 354 nm in CH_2Cl_2] were observed upon laser flash photolysis (308 nm excitation) of 46a under 1,2,4,5-tetracyanobenzene (TCNB, Scheme 11)- or *N*-methylquinolinium tetrafluoroborate ($NMQ^+BF_4^-$)-sensitized PET conditions. These transient absorption spectra resemble those of the 1,1-bis(4-methoxyphenyl)ethyl cation (48a$^+$, λ_{ab} = 484 nm in CH_3CN, Chart 1) and radical (48a$^•$, λ_{ab} = 350 nm in CH_3CN), respectively. These findings also suggest the twisted molecular geometry and localized electronic structure of 47a$^{•+}$ for the bis(4-methoxyphenyl)methyl cation subunit and the allyl radical subunit. It is also the same with 47a$^{•+}$, in which the bis(4-methoxyphenyl)methyl radical subunit was considerably twisted with the remaining allyl radical subunit. Density functional theory calculations support these conclusions.[17]

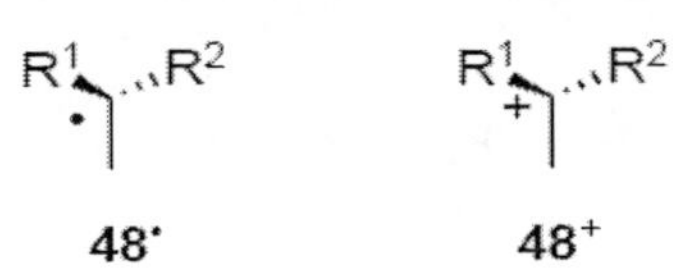

a: $R^1 = R^2$ = 4-$MeOC_6H_4$, **b**: $R^1 = R^2$ = Ph

Chart 1. The 1,1-diarylethyl radical (48$^•$) and cation (48$^+$) structurally related to 47$^{•+}$ and 47.

We also applied the time-resolved EPR technique to clarify the reaction sequence of the PET degenerate MCP rearrangement of 46a and succeeded in confirming the participation of $47a^{\bullet+}$. The chemically-induced dynamic electron polarization (CIDEP) spectrum of $47a^{\bullet+}$ was obtained at a delay time of 1 μs after the laser pulse (355 nm) under the chloranil (CA)-sensitized conditions in dimethyl sulfoxide at room temperature. The E*/E spectrum is easily reproduced by the superposition of triplet mechanism (E) and radical pair mechanism (E/A) signals,[18] where E and A denote emission and enhanced absorption of the microwave radiation, respectively. The $CA^{\bullet-}$ appears as the intense signal peak at 343.135 mT (g = 2.0058). The well-separated five lines are analyzed with two hyperfine splitting (hfs) constants corresponding to the $47a^{\bullet+}$ (g = 2.0031). Since the hfs constants and g-values shown in Chart 2 are close to those of the parent allyl radical, the unpaired electron is mainly distributed over the allyl subunit and the positive charge is localized on the bis(4-methoxyphenyl)methyl subunit of $47a^{\bullet+}$, indicating the twisted molecular geometry. These results are, thus, consistent with those elucidated by laser flash photolysis and CIDNP.

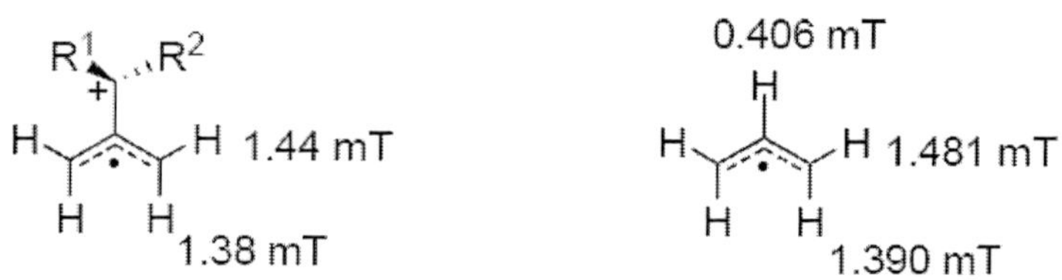

g = 2.0031 g = 2.0025

$47a^{\bullet+}$

a: $R^1 = R^2 = 4\text{-}MeOC_6H_4$

Chart 2. List of hfs constants and g-values of $47a^{\bullet+}$ and the parent ally radical.

A BET from $DCA^{\bullet-}$ to $47a^{\bullet+}$ is an important key process to generate another TMM-type intermediate, $^{3}47a^{\bullet\bullet}$. The most reliable way to prove the BET process may be to confirm a coincidence of a decay rate constant of $47a^{\bullet+}$ and a rise rate constant of $^{3}47a^{\bullet\bullet}$. Unfortunately, however, this kinetic study resulted in failure due to the difficulty in experimental analysis. We then studied the kinetics of the BET process theoretically. As described above, the interaction of bis(4-methoxyphenyl)methyl cation subunit and the allyl radical part in $47a^{\bullet+}$ is negligible, the reduction potential ($E^{red}_{1/2}$) of $47a^{\bullet+}$ is assumed to that of $48a^{+}$ (Chart 1), and thus to that of the oxidation potential ($E^{ox}_{1/2}$) of $48a^{\bullet}$, which is determined to be −0.06 V vs SCE in CH_3CN by photomodulation voltammetry.[15] Therefore, the free energy change is estimated to be ΔG_{BET} = $E^{ox}_{1/2}(DCA^{\bullet-}) - E^{red}_{1/2}(47a^{\bullet+}) \sim E^{red}_{1/2}(DCA) - E^{ox}_{1/2}(48a^{\bullet}) = -0.92$ eV in CH_3CN at 20 °C. Note that $E^{ox}_{1/2}$ of $DCA^{\bullet-}$ corresponds to $E^{red}_{1/2}$ of DCA (–0.98 V vs SCE). Calorimetric analysis using photoacoustic calorimetry and density functional theory (DFT) calculation also supports this conclusion.

The value of –0.92 eV, when used in Eqs. 1 and 2[19] based on the Marcus theory, enabled an estimate of the rate constant for BET (k_{BET}) from $DCA^{\bullet-}$ to $47a^{\bullet+}$ in a radical ion pair [$47a^{\bullet+}$, $DCA^{\bullet-}$]. The parameters in these equations were those suggested by Kikuchi and coworkers[20] (*i.e.,* the electronic coupling matrix element V, solvent reorganization energy λ_s, vibration reorganization energy λ_v, and single average frequency ν) and Planck's constant h, Boltzmann's constant k_b, and the temperature T, respectively. This calculation gave an estimated k_{BET} of 1.0 × 10^{10} s^{-1} in CH_3CN at 20 °C, suggesting efficient BET reaction to give $^{3}47a^{\bullet\bullet}$ and DCA.

$$k_{BET} = \left(\frac{4\pi^3}{h^2\lambda_s k_b T}\right)^{1/2} |V|^2 \sum_{w=0}^{\infty} \left(\frac{e^{-S} S^w}{w!}\right) \exp\left[-\frac{(\lambda_s + \Delta G_{BET} + wh\nu)^2}{4\lambda_s k_b T}\right] \quad (1)$$

$$S = \lambda_v / h\nu \quad (2)$$

5.4.2.2. PET Oxygenation of Arylated MCP

The intermediacy of the TMM-type radical cation $47a^{\bullet+}$ was also supported by oxygenation reaction. PET reaction of 46a using DCA under oxygen atmosphere gives 3,3-bis(4-methoxyphenyl)-4-methylene-1,2-dioxolane (49a, figure 5)[16] quantitatively as a major product in CH_3CN or CH_2Cl_2, but in a poor yield in the less polar solvent, benzene.[15] The TCNB-sensitized PET reactions in CH_3CN give similar results, though oxygenation in dichloromethane is less efficient. These facts strongly suggest that $47a^{\bullet+}$ intervenes in the PET degenerate MCP rearrangement of 46a. Therefore, under those sensitized conditions, time-resolved absorption spectroscopy on laser flash photolysis of 46a was performed.

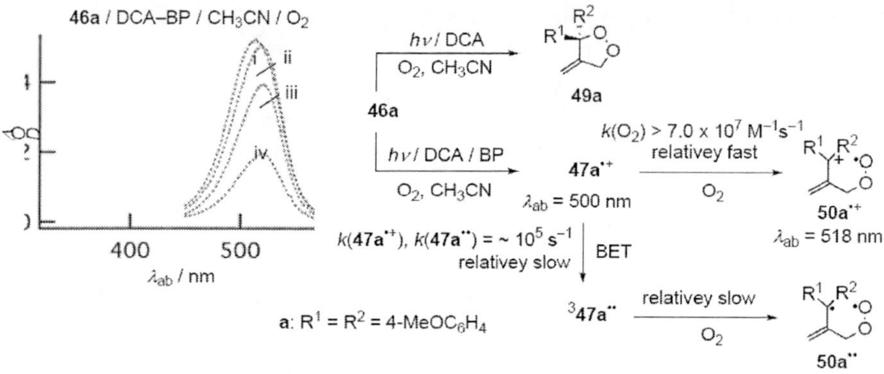

Figure 5. Time-resolved absorption spectra observed at a delay time of 60 ns (i), 200 ns (ii), 1 μs (iii), and 5 μs (iv) after laser flash photolysis (355 nm excitation) of 46a under DCA-sensitized PET conditions.

As described above, the decay kinetics of the 500- and 350-nm transients was assigned to $47a^{\bullet+}$ and $^{3}47a^{\bullet\bullet}$, respectively. The 500-nm transient observed with DCA as a sens and benzophenone (BP) as a co-sens persists for several μs under nitrogen in CH_3CN, but disappears completely within 200 ns under oxygen, whereby a new transient absorption band with λ_{ab} at 518 nm arises, as shown in figure 5. This transient is presumably due to a peroxide radical cation $50a^{\bullet+}$, leading to 49a. A second-order rate constant for the addition of molecular oxygen to free $47a^{\bullet+}$ [$k(O_2)$] is estimated to be at least 7.0×10^7 $M^{-1}s^{-1}$ in CH_3CN, while a pseudo first-order rate constant for the decay of $47a^{\bullet+}$ and for the rise of $^{3}47a^{\bullet\bullet}$ [$k(47a^{\bullet+})$ and $k(^{3}47a^{\bullet\bullet})$, but not a rate constant for BET between a $sens^{\bullet-}$ and a $47a^{\bullet+}$ in an ion radical pair] is assumed to be ~ 10^5 s^{-1}.[21] Furthermore, judging from the thermolysis of 46a under oxygen, the addition of molecular oxygen to $^{3}47a^{\bullet\bullet}$ does not seem to be very fast. This is in line with the results of laser flash photolysis. Thus, the key process of PET oxygenation of 46a is probably a capture of $47a^{\bullet+}$ with molecular oxygen, but not of $^{3}47a^{\bullet\bullet}$. As described above, no significant interaction is suggested between the two chromophores [the bis(4-methoxyphenyl)methyl cation or radical subunit and the allyl radical subunit] in both $47a^{\bullet+}$ and $^{3}47a^{\bullet\bullet}$. Interestingly, however, a possible difference in reactivity to molecular oxygen between $47a^{\bullet+}$ and $^{3}47a^{\bullet\bullet}$ implies a possible communication between the two chromophores.

5.4.3. Thermoluminescence (TL) and Organic Radical Light-Emitting Diode (ORLED) Using Arylated MCP

In the last section, we described a full mechanism for the PET degenerate rearrangement of arylated MCP (46, Scheme 14) that involves a facile BET to convert $47^{\bullet+}$ to $^{3}47^{\bullet\bullet}$. A similar rearrangement can also be triggered by γ-irradiation or electrically.[22] We have recently demonstrated thermoluminescence (TL) of 46 using γ-irradiation–annealing technique and applied this TL phenomena to novel organic light-emitting diodes (OLEDs) utilizing highly reactive open-shell molecules, such as $^{3}47^{\bullet\bullet}*$, as the light-emitting species.[22] We call this new device an "organic *radical* light-emitting diode (ORLED)".[24] The ORLED approach has the potential of overcoming major problems associated with typical OLED systems, such as difficulties in obtaining long wavelength emissions, low quantum efficiency, and low device durability. Here, TL and ORLEDs using 46 are described briefly.

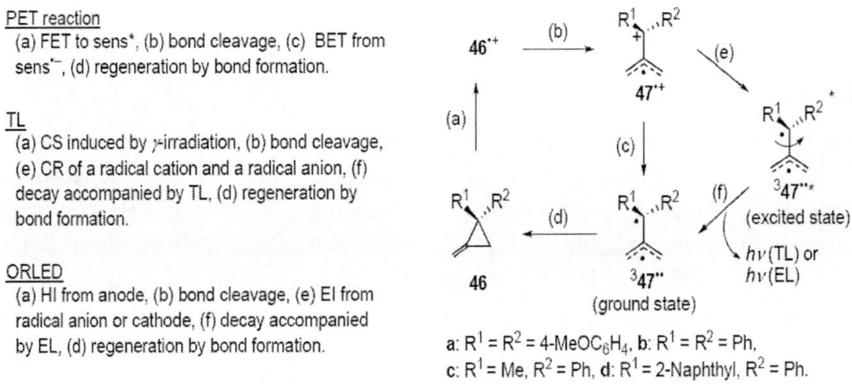

Scheme 14. A mechanistic connection among the PET reaction, TL, and ORLED of 46.

5.4.3.1. TL Using Arylated MCP

TL refers to light emission triggered by thermal activation. TL usually occurs when an electronically excited singlet state is formed by charge recombination (CR, corresponding to BET) of a radical cation with a radical anion. A typical example in chemistry is the light emission from a substrate triggered by the annealing of a γ-irradiated, low-temperature glassy matrix. Similar to PET reaction pathways described above, the TL procedure involves charge separation (CS, corresponding to FET) and CR as key mechanistic steps. The PET and TL systems exhibit one major energetic difference. For example in the PET reaction of 46, BET to $47^{\bullet+}$ from $sens^{\bullet-}$ produces a ground state $^{3}47^{\bullet\bullet}$. In TL systems, CR of $47^{\bullet+}$ and $46^{\bullet-}$ leads to the formation of the excited state $^{3}47^{\bullet\bullet}*$. This contrast originates from the difference in $E^{red}_{1/2}$ between sens and 46.

The degassed methylcyclohexane (MCH) glassy matrix containing 46b (R^1 = R^2 = Ph, 5 mM) was γ-irradiated at 77 K using a ^{60}Co source. MCH is a solvent that is generally employed in γ-irradiation studies to simultaneously generate radical cations and anions. A matrix prepared in this way contains $46b^{\bullet+}$, $47b^{\bullet+}$, $46b^{\bullet-}$, and $47b^{\bullet-}$ (Chart 3). The absorbance maxima at 440 and 500 nm, observed at 77 K, were assigned to $47b^{\bullet+}$ and $47b^{\bullet-}$, respectively.[24]

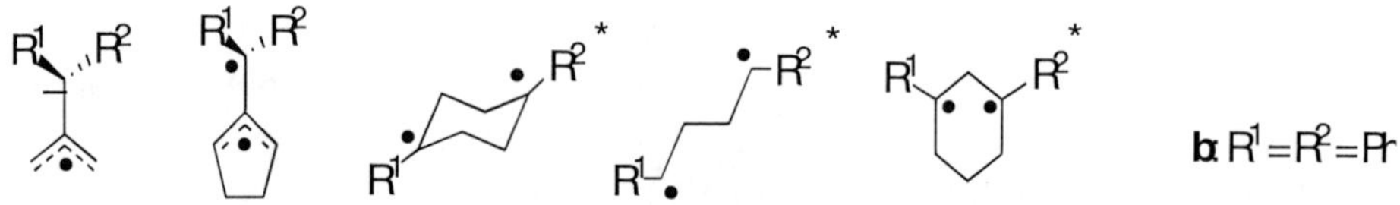

$47b^{\bullet-}$ $^{3}51b^{\bullet\bullet}$ $^{1}52b^{\bullet\bullet}*$ $^{1}53b^{\bullet\bullet}*$ $54b^{\bullet\bullet}*$

Chart 3. A list of key intermediates.

Upon annealing the MCH matrix from 77 to 130 K, an intense bright green TL was observed [figure 6 (center), λ_{tl} = 501 and 533 nm], and assigned to $^{3}47b^{\bullet\bullet}*$. To our knowledge, this is the first observation of TL from an electronically excited biradical species. The excited TMM biradical $^{3}47b^{\bullet\bullet}*$ has the considerably twisted molecular geometry and the localized electronic state, similar to $47b^{\bullet+}$ and $^{3}47b^{\bullet\bullet}$,[15,17] because TL band of $^{3}47b^{\bullet\bullet}*$ resembled to the photoluminescence band of the structurally similar 1,1-diphenylethyl radical ($48b^{\bullet}$, Chart 1) and diphenyl-substituted, five-membered cyclic TMM biradical ($^{3}51b^{\bullet\bullet}$, Chart 3).[25] Similarly, blue [figure 6 (left), λ_{tl} = 462 and 484 nm] and red [figure 6 (right), λ_{tl} = 603 and 656 nm] TL was observed when MCH matrices of 46c (R^1 = Me, R^2 = Ph, Scheme 14) and 46d (R^1 = 2-Naphthyl, R^2 = Ph) were independently γ-irradiated. They are assigned to $^{3}47c^{\bullet\bullet}*$ and $^{3}47d^{\bullet\bullet}*$, respectively.

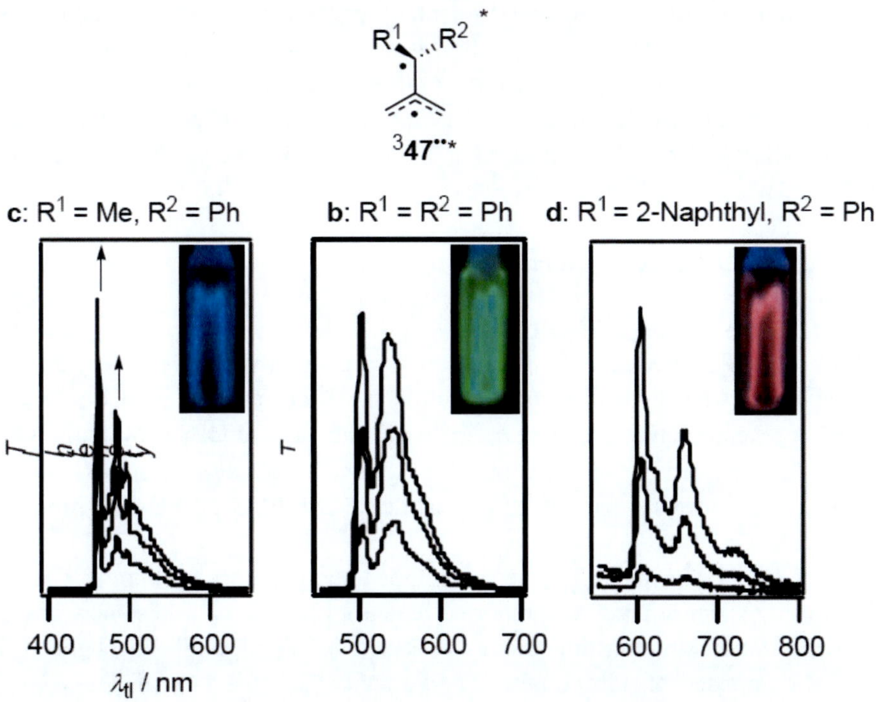

Figure 6. Photos and changes in the TL spectra of $^{3}47b^{\bullet\bullet}*$ (center, green), $^{3}47c^{\bullet\bullet}*$ (left, blue), and $^{3}47d^{\bullet\bullet}*$ (right, red) observed on warming the γ-irradiated glassy MCH matrices containing 46b, 46c, and 46d respectively, from 77 to *ca.* 130 K. Arrows indicate the increase in the intensities of the emission bands.

A plausible mechanism for the TL of 46b consists of five steps (Scheme 14): (a) CS induced by γ-irradiation to give $46b^{\bullet+}$ and $46b^{\bullet-}$.[24] (b) bond cleavage of $46b^{\bullet+}$ or $46b^{\bullet-}$ to give $47b^{\bullet+}$ or $47b^{\bullet-}$, respectively,[15] (e) CR from a radical anion to give $^{3}47b^{\bullet\bullet}*$, (f) decay of $^{3}47b^{\bullet\bullet}*$ to $^{3}47b^{\bullet\bullet}$ accompanied by TL, and (d) quantitative regeneration of 46b by bond formation in $^{3}47b^{\bullet\bullet}$. Note that the ground state of $47b^{\bullet\bullet}$ exhibits triplet multiplicity, judging from the Curie plot analysis and DFT calculations.[15,22] The excited TMM biradical $47b^{\bullet\bullet}*$ is likely in the triplet state ($^{3}47b^{\bullet\bullet}*$), and thus TL in the 46b system would correspond to *triplet–triplet fluorescence* from $^{3}47b^{\bullet\bullet}*$. Since TL associated with the decay of $^{3}47b^{\bullet\bullet}*$ resembles the photoluminescence of $48b^{\bullet}$, the emission from $^{3}47b^{\bullet\bullet}*$ most likely occurs primarily from the diphenylmethyl radical fluorophore.[25] The remaining allyl subunit is not involved in the fluorescence emission, but it plays an important role in the bond-forming process leading back to the formation of 46b.

A similar TL was also observed for many organic radicals. For example, we have observed TL of the singlet excited states of 1,4-diphenylcyclohexane-1,4-diyl (^{1}52b$^{\bullet\bullet}$*, Chart 3)[26] and 1,4-diphenylbutane-1,4-diyl (^{1}53b$^{\bullet\bullet}$*) at approximately 586 and 520 nm in MCH, respectively. 1,3-Diphenylcyclohexane-1,3-diyl (54b$^{\bullet\bullet}$*) also exhibited similar TL at approximately 572 nm while its multiplicity has not been elucidated yet.

5.4.3.2. ORLED Using Arylated MCP

5.4.3.2.1. Basic Concept of ORLED

Like the TL mechanism, OLED emission involves hole injection (HI, corresponding CS) and electron injection (EI, corresponding CR) as key mechanistic steps. A major difference between the TL and OLED systems is that CS in TL is triggered by γ-irradiation of the glassy matrix at low temperature and CR is induced by subsequent annealing, while HI and EI in OLEDs are promoted electronically.

Novel OLEDs using highly reactive, open-shell molecules like ^{3}47$^{\bullet\bullet}$ as the light-emitting species have been reported recently.[22] We call these new devices "organic *radical* light-emitting diodes (ORLEDs)." A plausible mechanism for ORLEDs composed of 46b consists of five steps (Scheme 14): (a) HI from the anode to produce 46b$^{\bullet+}$, (b) bond cleavage of 46b$^{\bullet+}$ to give 47b$^{\bullet+}$, (e) EI from the radical anion or cathode to 46b$^{\bullet+}$ to give ^{3}47b$^{\bullet\bullet}$*, (f) decay of ^{3}47b$^{\bullet\bullet}$* to ^{3}47b$^{\bullet\bullet}$ accompanied by electroluminescence (EL), and (d) quantitative regeneration of 46b by bond formation in ^{3}47b$^{\bullet\bullet}$. Design strategies of ORLED systems may seem impractical due to the high durability required in these devices. However, ORLEDs have several potentially advantageous characteristics: long wavelength EL is more easily generated with easily prepared, smaller molecules; the internal EL quantum efficiency (η_{int}) is significantly higher; and regeneration of the starting material is relatively more efficient (figure 7).

Long wavelength EL from an open-shell molecule (organic radical) arises from the HOMO–SOMO or SOMO–LUMO energy gaps, which are generally much smaller than the HOMO–LUMO gaps in structurally similar closed-shell molecules. This is illustrated in figure 7 and was confirmed by the observation of green TL from ^{3}47b$^{\bullet\bullet}$*.[22]

Based on spin multiplicity, we expected ORLED systems based on ^{3}47$^{\bullet\bullet}$ to have a large η_{int}. A typical OLED, based on the fluorescence of a closed-shell molecule ($^{1}S_a$, figure 7), has a maximum η_{int} of 25% because $^{1}S_a$* is converted from a $S_a^{\bullet+}$ hole in a statistically-predicted 25% yield; the fluorescence takes place as a singlet–singlet transition (figure 7). The triplet excited state ($^{3}S_a$*) is generated simultaneously in 75% yield, and undergoes spin-forbidden and nonradiative decays. This is energy wasting deactivation of the excited species. In contrast, the triplet excited state ($^3I^{\bullet\bullet}*$) of open-shell molecules (intermediates) such as ^{3}47$^{\bullet\bullet}$, that are generated in 75% yield from $S_b^{\bullet+}$ together with the 25% singlet excited state, can undergo *triplet–triplet transitions* to produce triplet ground states ($^3I^{\bullet\bullet}$) with highly efficient fluorescence emission. As a result, ORLEDs based on triplet species such as ^{3}47$^{\bullet\bullet}$ can reach theoretically maximal η_{int} values of 75%.

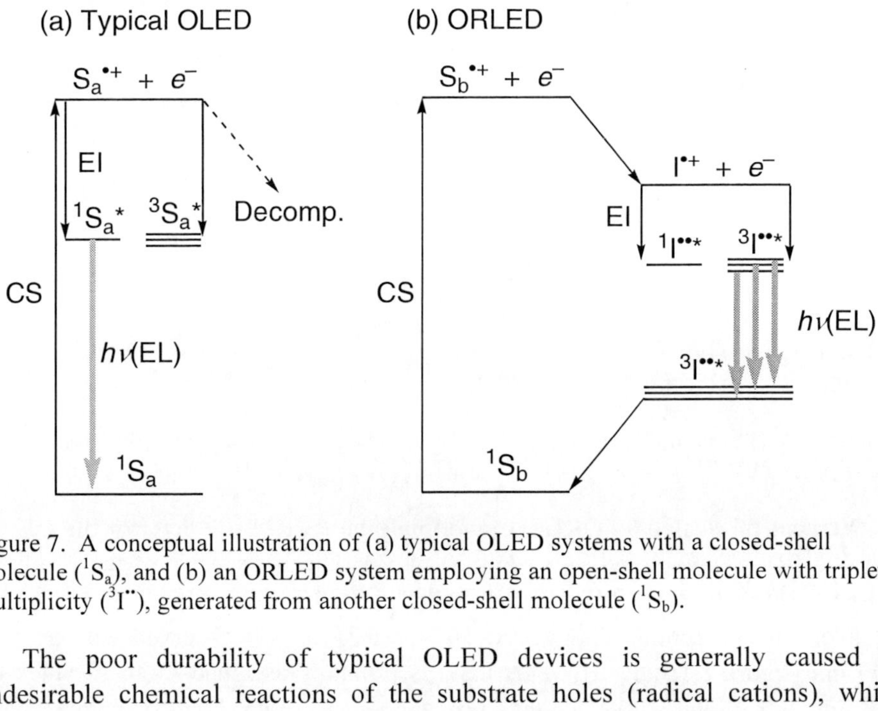

Figure 7. A conceptual illustration of (a) typical OLED systems with a closed-shell molecule (^1S_a), and (b) an ORLED system employing an open-shell molecule with triplet multiplicity ($^3I^{\bullet\bullet}$), generated from another closed-shell molecule (^1S_b).

The poor durability of typical OLED devices is generally caused by undesirable chemical reactions of the substrate holes (radical cations), which resulted in the decomposition. This problem may be overcome by using open-shell molecules that allow efficient regeneration of the starting material (figure 7). For example, 47$^{\bullet+}$ is converted to ^{3}47$^{\bullet\bullet}$*, which may be quantitatively regenerated to 46 via ^{3}47$^{\bullet\bullet}$ after light emission (figure 6).[22]

5.4.3.2.2. Trial Manufacture of ORLED Using Arylated MCP

We first manufactured a bilayer OLED with an indium tin oxide (ITO) anode layer coated with a layer of poly(3,4-ethylenedioxythiophene) (PEDOT)–polystyrene sulfonate (PSS) as the hole-injection/transport material [figure 8 (left)]. Poly(*N*-vinylcarbazole) (PVK) and 2-(4-biphenylyl)-5-(4-*t*-butylphenyl)-

1,3,4-oxadiazole (PBD) were deposited as host materials, and topped with a Ca–Al cathode {[ITO/PEDOT–PSS/PVK–PBD (67:13)/Ca–Al]}. A 15-V voltage applied to the OLED produced an EL band with a maximum (λ_{el}) at 458 nm, corresponding to emission from PVK* [figure 8 (right, i)].

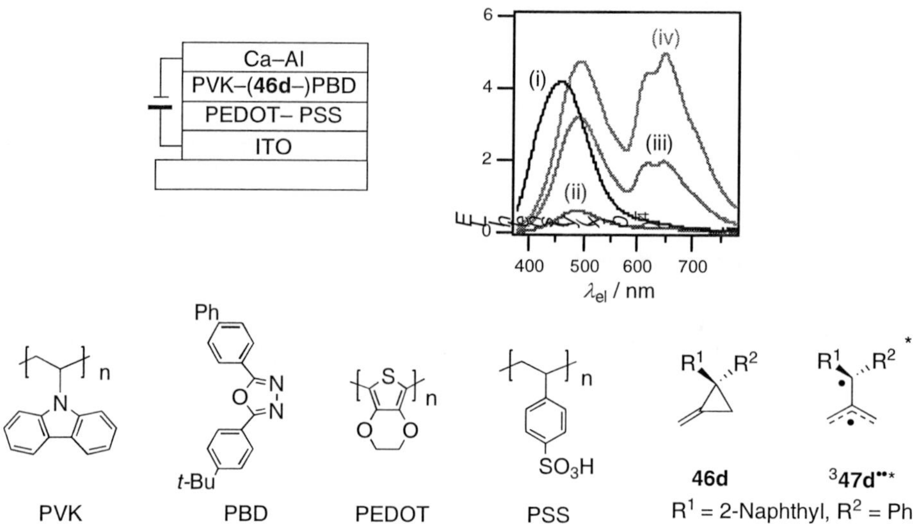

Figure 8. (Left) Structure of the bilayer ORLED with or without dopant 46d. (Right) EL spectra of the ORLED without [(i) 15 V] or with [(ii) 10 V, (iii) 15 V, (iv) 20 V] 46d.

We next fabricated an ORLED by doping the PVK–PBD layer of the OLED with 46d (R^1 = 2-Naphthyl, R^2 = Ph, 20 wt.%). The EL spectra of the resulting ORLED [ITO/PEDOT–PSS/PVK–46d–PBD(67:20:13)/Ca–Al] were strongly dependent on the applied voltage. At 10 V, weak EL was observed with a λ_{el} at ~490 nm [figure 8 (right, ii)]. This band is probably assigned to an exciplex of PVK and 46d. When a 20-V voltage was applied, EL with λ_{el} at 617 and 650 nm appeared together with λ_{el} at 493 nm [figure 8 (right, iv)]. The EL bands with λ_{el} at 617 and 650 nm corresponds to the 603- and 656-nm TL bands observed for the MCH matrix of 46d. This implies the formation of ^{3}47d$^{\bullet\bullet}$* at ambient temperature in the ORLED, which then acted as the light-emitting species.

In its current form, the ORLED containing 46d is limited by high voltage requirements (~20 V), low luminescence intensity (~30.0 cd m^{-1}), poor current efficiency (~0.158 cd A^{-1}), and still low device durability. This may be due to undesired reactions of 47d$^{\bullet+}$ or 47d$^{\bullet\bullet}$* or 47d$^{\bullet\bullet}$ with molecular oxygen (described above) or some other reactants. In spite of these shortcomings, ORLED designs based on chemical reactions of 46 and light emission from short-lived open-shell molecules $^{3}47d^{\cdot\cdot *}$ may prove viable for the construction of highly efficient devices.

The expected and observed mechanistic features of the ORLED have the potential of overcoming major current problems such as difficulty in obtaining long wavelength emission, low quantum efficiency, and low device durability, which are associated with typical OLEDs that rely on organic, closed-shell hydrocarbons. Especially, device durability is important in practical use. Therefore, the use of a reactive species such as $^{3}47^{\cdot\cdot *}$ would appear to be a paradoxical approach. While it is currently not clear whether the ORLED system described here can be put to practical use, this paradoxical approach has created a new category of luminescence systems.

5.5. Highly Emissive Organosilicon Compounds

Organosilicon compounds are environmentally-friendly and most of them can be easily synthesized from commercially available halosilanes. The photochemistry of organosilicon compounds[27] has been developed with regard to the absorption and emission of polysilanes,[28] generation of silylene and disilene by photoirradiation to oligosilanes,[29] photochemistry of intramolecular charge transfer complexes of aromatic disilanes,[30] PET reactions of organosilicon compounds.[9] Recently, the fluorescence increasing effect of aromatic hydrocarbons such as anthracenes, naphthacenes, pentacenes, triphenylenes by introduction of silyl groups are discussed in detail.[31] We studied effects of silyl, germyl, and stannyl groups on absorption and fluorescence properties of pyrene.[32]

Mono-, di-, tri-, and tetrabromopyrenes can be synthesized by the reaction of pyrene (20) with appropriate equivalents of bromine. Trimethylsilyl-, trimethylgermyl-, and trimethylstannyl-substituted pyrenes 55-60 were prepared by lithiation of these bromopyrenes and the following treatment with Me_3MCl (M = Si, Ge, Sn). Trimethylsilylethynyl-substituted pyrenes 61-64 were synthesized by Sonogashira reaction of bromopyrenes with trimethylsilylacetylene.

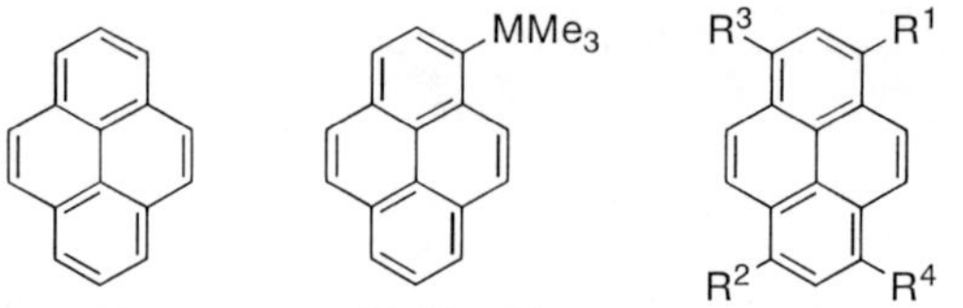

20

55: M = Si
56: M = Ge
57: M = Sn

55: R^1 = $SiMe_3$, R^2 = R^3 = R^4 = H
58: R^1 = R^2 = $SiMe_3$, R^3 = R^4 = H
59: R^1 = R^2 = R^3 = $SiMe_3$, R^4 = H
60: R^1 = R^2 = R^3 = R^4 = $SiMe_3$

61: R^1 = C≡C−$SiMe_3$, R^2 = R^3 = R^4 = H
62: R^1 = R^2 = C≡C−$SiMe_3$, R^3 = R^4 = H
63: R^1 = R^2 = R^3 = C≡C−$SiMe_3$, R^4 = H
64: R^1 = R^2 = R^3 = R^4 = C≡C−$SiMe_3$

65: R^1 = R^2 = R^3 = R^4 = C≡C−Bu-*t*

Chart 4. Silyl-, germyl-, and stannyl-substituted pyrene derivatives and the related compounds.

The UV-vis absorption and fluorescence spectra of these compounds were measured in cyclohexane (ca. 10^{-5} M). Absorption maxima of trimethylsilyl-, trimethylgermyl-, and trimethylstannyl-substituted pyrenes (55-57) shifted to bathochromic region than that of unsubstituted pyrene (20), although the absorption coefficients were almost unchanged. The absorption maxima of mono-, bis-, tris-, and tetrakis(trimethylsilyl)pyrenes (55, 58-60) shifted to longer wavelength consecutively at intervals of 10 nm. These results can be explained in terms of the energy gap between HOMO and LUMO levels of pyrene derivatives, which decreased with increasing the silyl-substitution. Fluorescence intensity and fluorescence lifetime of the monosilyl derivative (55) were higher and longer than those of 20. Fluorescence intensities and lifetimes of the germyl and stannyl derivatives 56 and 57 were lower and shorter than those of 20 due to the heavy atom effect. It is interesting that the fluorescence intensities increased with increasing the number of silyl substitution (i.e. 55 < 58 < 59 < 60). However, fluorescence lifetimes decreased with increasing the number of silyl groups. These phenomena are also observed in the absorption and fluorescence properties of silylethynylpyrenes 61-64.[33] The absorption maxima of 61-64 consecutively shifted to longer wavelengths, and their molar extinction coefficients increase with increasing number of trimethylsilylethynyl groups. Fluorescence quantum yields (Φ_f) of 61-64 are essentially high and increase with increasing number of silylethynyl groups. Particularly, Φ_f for 64 is >0.99, meaning that the absorbed photons by 64 are exclusively emitted as fluorescence (figure 9). Conversely, the fluorescence lifetimes of 61-64 decrease with increasing the number of attached silylethynyl groups. Silylethynylpyrenes 61-64 showed excimer emissions in the wavelength region of 500-570 nm at high concentrations in CH_2Cl_2.

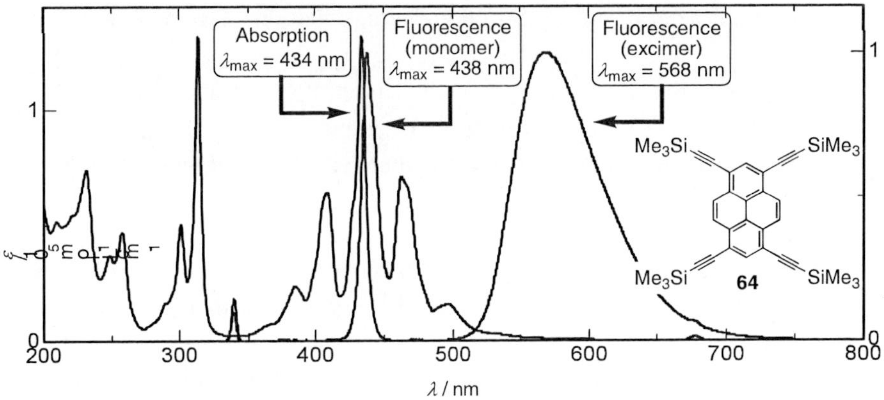

Figure 9. Absorption and fluorescence spectra of tetrakis(trimethylsilylethynyl)pyrene (64).

The main process from the excited molecules is figured in Jablonski diagram (figure 10). The excited singlet state (S_1), generated by photoexcitation of the ground-state molecule (S_0), decays with fluorescence radiation, nonradiative internal conversion, intersystem crossing to the excited triplet state (T_1), and chemical reactions. From T_1, decay paths with phosphorescence radiation, nonradiative internal conversion, and chemical reactions are involved. Fluorescence quantum yield Φ_f can be described as a product of the rate constant for fluorescence radiation (k_f) and fluorescence lifetime (τ_f), i.e. $\Phi_f = k_f\tau_f$. Although the reason why silyl and silylethynyl substituents can increase the fluorescence quantum yield is not fully understood, we suggest the following effects.

1. The absorption spectrum of 1,3,6,8-tetrakis(*tert*-butylethynyl)pyrene (65), the corresponding carbon counterpart of 64, is less shifted bathochromically compared with that of 64. The effect of the silicon atoms on the extra bathochromic shift might be due to σ-π interaction between the C-Si(σ) and the acetylenic π bonds. This σ-π interaction is also supported by B3LYP/6-31G(d) calculation in which both HOMO and LUMO spread in pyrene π framework and σ(Si-Me) bonds. The σ-π interaction acts to extend the chromophore.
2. When the difference in structures between the ground state and excited state is small (i.e., Stokes shift is small), the overlap of wave functions of the ground state and excited state becomes large. Indeed, as expected from the rigid structures of the alkynylpyrenes, their Stokes shifts are

essentially small. From Strickler-Berg equation, $1/\tau_f = 3 \times 10^{-9} n^2 \nu_m^{-2} \int \varepsilon(\nu) d\nu$, where n is the refractive index of solvent, ν_m^{-2} is the energy of the absorption maximum and the integration is the area under the absorption curve, large transition moment causes large ε and large k_f values.[34] It has been discussed that silyl substituents increase the transition moments when they are directly attached to thiophenes.[35]

3. The energy level of T_2 of the parent anthracene is somewhat lower than that of S_1. However, when trialkylsilyl groups are introduced to 9,10-positions of anthracene, T_2 level becomes higher than S_1 level, so the decay path of S_1 through T_2 is diminished when silyl groups are introduced to anthracene.[36] In the pyrene derivatives, energy gap between S_1 and T_2 may also be involved.

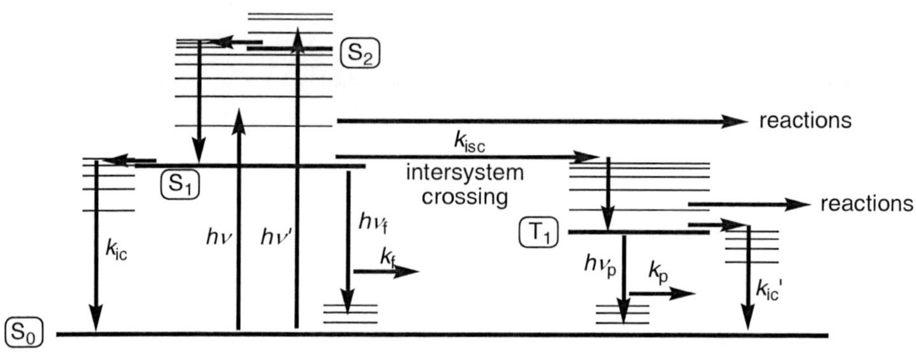

Figure 10. Jablonski diagram that explains the main process in photochemistry.

CONCLUSIONS

In green chemistry, efficient and effective environmentally harmonious organic reactions are required. In this chapter, a variety of organic photochemical reactions are described from synthetic and mechanistic viewpoints. We have found not only new types of photocycloaddition and photoaddition reactions, but also the quite interesting phenomena of thermoluminescence. We, thus, believe strongly that organic photochemistry will contribute greatly to atom-economically and eco-friendly synthetic chemistry.

References

[1] (a) McCullough, J. J. *Chem. Rev.* 1987, *87*, 811–860. (b) Mizuno, K.; Maeda, H.; Sugimoto, A.; Chiyonobu, K. in *Understanding and Manipulating Excited-State Processes*, Ramamurthy, V.;Schanze, K. S. eds., Marcel Dekker: New York, 2001, 127–241. (c) Hoffmann, N. in *Synthetic Organic Photochemistry*, Griesbeck, A. G.; Mattay, J. eds., Marcel Dekker, New York, 2005, 529–552.

[2] (a) Mizuno, K.; Pac, C.; Sakurai, H. *J. Chem. Soc., Chem. Commun.* 1974, 648–649. (b) Mizuno, K.; Takenaka, H.; Maeda, H, unpublished results. (c) Yokoyama, A.; Mizuno, K. *Org. Lett.* 2000, *2*, 3457–3459. (d) Chiyonobu, K.; Maeda, H.; Mizuno, K. unpublished results.

[3] (a) Caldwell, R. A.; Creed, D. *Acc. Chem. Res.* 1980, *13*, 45–50. (b) Mizuno, K.; Caldwell, R. A.; Tachibana, A.; Otsuji, Y. *Tetrahedron Lett.* 1992, *33*, 5779–5782. (c) Mizuno, K.; Maeda, H.; Inoue, Y.; Sugimoto, A.; Vo, L. P.; Caldwell, R. A. *Tetrahedron Lett.* 2000, *41*, 4913–4916. (d) Maeda, H.; Sugimoto, A.; Mizuno, K. *Org. Lett.* 2000, *2*, 3305–3308.

[4] Mukae, H.; Maeda, H.; Mizuno, K. *Angew. Chem. Int. Ed.* 2006, *45*, 6558–6560.

[5] (a) Lu, H.; Schmidt, M. A.; Jensen, K. F. *Lab Chip* 2001, *1*, 22–28. (b) Ueno, K.; Kitagawa, F.; Kitamura, N. *Lab Chip* 2002, *2*, 231–234. (c) Fukuyama, T.; Hino, Y.; Kamata, N.; Ryu, I. *Chem. Lett.* 2004, *33*, 1430–1431. (d) Matsushita, Y.; Kumada, S.; Wakabayashi, K.; Sakeda, K.; Ichimura, T. *Chem. Lett.* 2006, *35*, 410–411. (e) Kitamura, N. *Chem. Educ.* 2002, *50*, 678–681. (f) Nagaki, A.; Togai, M.; Suga, S.; Aoki, N.; Mae, K.; Yoshida, J. *J. Am. Chem. Soc.* 2005, *127*, 11666–11675.

[6] (a) Maeda, H.; Mukae, H.; Mizuno, K. *Chem. Lett.* 2005, *34*, 66–67. (b) Mukae, H.; Maeda, H.; Nashihara, S.; Mizuno, K. *Bull. Chem. Soc. Jpn.* 2007, *80*, 1157–1161.

[7] (a) McCullough, J. J.; MacInnis, W. K.; Lock, C. J. L.; Faggiani, R. *J. Am. Chem. Soc.* 1980, *102*, 7780–7782. (b) McCullough, J. J.; MacInnis, W. K.; Lock, C. J. L.; Faggiani, R. *J. Am. Chem. Soc.* 1982, *104*, 4644–4658.

[8] Mizuno, K.; Otsuji, Y. *Top. Curr. Chem.* 1994, *169*, 301–346.

[9] Mizuno, K.; Tamai, T.; Sugimoto, A.; Maeda, H. in *Advances in Electron Transfer Chemistry*, Mariano, P. S., ed., JAI Press: London, 1999, Vol. 6, 131–165.

[10] (a) Mizuno, K.; Ikeda, M.; Otsuji, Y. *Tetrahedron Lett.* 1985, *26*, 461–464. (b) Nakanishi, K.; Mizuno, K.; Otsuji, Y. *Bull. Chem. Soc. Jpn.* 1993, *66*, 2371–2379. (c) Nishiyama, T.; Mizuno, K.; Otsuji, Y.; Inoue, H.

Tetrahedron, 1995, *51*, 6695–6706. (d) Mizuno, K.; Ikeda, M.; Otsuji, Y. *Chem. Lett.* 1988, 1507–1510. (e) Hayamizu, T.; Ikeda, M.; Maeda, H.; Mizuno, K. *Org. Lett.* 2001, *3*, 1277–1280.

[11] (a) Ohashi, M.; Nakatani, K.; Maeda, H.; Mizuno, K. *Org. Lett.* 2008, *10*, 2741–2743. (b) Ohashi, M.; Nakatani, K.; Maeda, H.; Mizuno, K. *J. Org. Chem.* 2008, *73*, 8348–8351.

[12] (a) Berson, J. A. In *Rearrangements in Ground and Excited States*, Mayo, P. d.; Ed.; Organic Chemistry, A Series of Monographs 42, Academic: New York, NY, 1980; Vol. 1, pp 311–390. (b) Gajewski, J. J. In *Hydrocarbon Thermal Isomerizations*. Wasserman, H. H. Ed.; Organic Chemistry, A Series of Monographs 44, Academic: New York, 1981; Vol. 5, pp 43–70.

[13] Gilbert, J. C.; Butler, J. R. *J. Am. Chem. Soc.* 1970, *92*, 7493–7495.

[14] Miyashi, T.; Takahashi, Y.; Mukai, T.; Roth, H. D.; Schilling, M. L. M. *J. Am. Chem. Soc.* 1985, *107*, 1079–1080.

[15] (a) Ikeda, H.; Nakamura, T.; Miyashi, T.; Goodman, J. L.; Akiyama, K.; Tero-Kubota, S.; Houmam, A.; Wayner, D. D. M. *J. Am. Chem. Soc.* 1998, *120*, 5832–5833. (b) Ikeda, H.; Akiyama, K.; Takahashi, Y.; Nakamura, T.; Ishizaki, S.; Shiratori, Y.; Ohaku, H.; Goodman, J. L.; Houmam, A.; Wayner, D. D. M.; Tero-Kubota, S.; Miyashi, T. *J. Am. Chem. Soc.* 2003, *125*, 9147–9157.

[16] Takahashi, Y.; Miyashi, T.; Mukai, T. *J. Am. Chem. Soc.* 1983, *105*, 6511–6513.

[17] Namai, H.; Ikeda, H.; Kato, N.; Mizuno, K. *J. Phys. Chem. A* 2007, *111*, 4436–4442.

[18] McLauchlan, K. A. In *Modern Pulsed and Continuous-Wave Electron Spin Resonance*; Keva, L., Bowman, M. K., Eds.; Wiley: New York, 1990, p 285–363.

[19] Miller, J. R.; Beitz, J. V.; Huddleston, R. K. *J. Am. Chem. Soc.* 1984, *106*, 5057–5068.

[20] Niwa, T.; Kikuchi, K.; Matsusita, N.; Hayashi, M.; Katagiri, T.; Takahashi, Y.; Miyashi, T. *J. Phys. Chem.* 1993, *97*, 11960–11964.

[21] Ikeda, H.; Ikeda, T.; Akagi, M.; Namai, H.; Miyashi, T.; Takahashi, Y.; Kamata, M. *Tetrahedron Lett.* 2005, *46*, 1831–1835.

[22] Namai, H.; Ikeda, H.; Hoshi, Y.; Kato, N.; Morishita, Y.; Mizuno, K. *J. Am. Chem. Soc.* 2007, *129*, 9032–9036.

[23] Ikeda, H. J. Photopolym. Sci. Technol. 2008, 21, 327–332.

[24] (a) Ikeda, H.; Namai, H.; Kato, N.; Ikeda, T. *Tetrahedron Lett.* 2006, *47*, 1857–1860. (b) Ikeda, H.; Namai, H.; Kato, N.; Ikeda, T. *Tetrahedron Lett.* 2006, *47*, 1501–1504.

[25] (a) Turro, N. J.; Mirbach, M. J.; Harrit, N.; Berson, J. A.; Platz, M. S. *J. Am. Chem. Soc.* 1978, *100*, 7653–7658. (b) Ikeda, H.; Namai, H.; Taki, H.; Miyashi, T. *J. Org. Chem.* 2005, *70*, 3806–3813.

[26] Namai, H.; Ikeda, H.; Hoshi, Y.; Mizuno, K. *Angew. Chem., Int. Ed.* 2007, *46*, 7396–7398.

[27] (a) Steinmetz, M. G. *Chem. Rev.* 1995, *95*, 1527–1588. (b) Bock, H.; Solouki, B. *Chem. Rev.* 1995, *95*, 1161–1190.

[28] Miller, R. D.; Michl, J. *Chem. Rev.* 1989, *89*, 1359–1410.

[29] (a) West, R.; Fink, M. J.; Michl, J. *Science* 1981, *214*, 1343–1344. (b) Murakami, S.; Collins, S.; Masamune, S. *Tetrahedron Lett.* 1984, *25*, 2131–2134. (c) Michalczyk, M. J.; West, R.; Michl, J. *Organometallics* 1985, *4*, 826–829. (d) Dewan, J. C.; Murakami, S.; Snow, J. T.; Collins, S.; Masamune, S. *J. Chem. Soc., Chem. Commun.* 1985, 892–894. (e) Yokelson, H. B.; Maxka, J.; Siegel, D. A.; West, R. *J. Am. Chem. Soc.* 1986, *108*, 4239–4241. (f) Shepherd, B. D.; Powell, D. R.; West, R. *Organometallics* 1989, *8*, 2664–2669.

[30] (a) Shizuka, H.; Sato, Y.; Ueki, Y.; Ishikawa, M.; Kumada, M. *J. Chem. Soc., Faraday Trans. 1*, 1984, *80*, 341–357. (b) Sakurai, H. *Pure Appl. Chem.* 1987, *59*, 1637–1646.

[31] (a) Kyushin, S.; Ikarugi, M.; Goto, M.; Hiratsuka, H.; Matsumoto, H. *Organometallics* 1996, *15*, 1067–1070. (b) Yamaguchi, S.; Akiyama, S.; Tamao, K. *Organometallics* 1998, *17*, 4347–4352. (c) Karatsu, T.; Shibata, T.; Nishigaki, A.; Kitamura, A.; Hatanaka, Y.; Nishimura, Y.; Sato, S.; Yamazaki, I. *J. Phys. Chem. B* 2003, *107*, 12184–12191. (d) Karatsu, T.; Terasawa, M.; Yagai, S.; Kitamura, A.; Nakamura, T.; Nishimura, Y.; Yamazaki, I. *J. Organomet. Chem.* 2004, *689*, 1029–1035. (e) Karatsu, T.; Hazuku, R.; Nishioka, D.; Yagai, S.; Kobayashi, N.; Kitamura, A. *J. Photopolym. Sci. Tech.* 2005, *18*, 65–68. (f) Kyushin, S.; Ishikita, Y.; Matsumoto, H.; Horiuchi, H.; Hiratsuka, H. *Chem. Lett.* 2006, *35*, 64–65. (g) Anthony, J. E.; Brooks, J. S.; Eaton, D. L. Parkin, S. R. *J. Am. Chem. Soc.* 2001, *123*, 9482–9483. (h) Bénard, C. P.; Geng, Z.; Heuft, M. A.; VanCrey, K.; Fallis, A. G. *J. Org. Chem.* 2007, *72*, 7229–7236. (i) Kyushin, S.; Takemasa, N.; Matsumoto, H.; Horiuchi, H.; Hiratsuka, H. *Chem. Lett.* 2003, *32*, 1048–1049. (j) Declercq, D.; Delbeke, P.; De Schryver, F. C.; Van Meervelt, L.; Miller, R. D. *J. Am. Chem. Soc.* 1993, *115*, 5702–5708.

[32] Maeda, H.; Inoue, Y.; Ishida, H.; Mizuno, K. *Chem. Lett.* 2001, 1224–1225.

[33] (a) Maeda, H.; Maeda, T.; Mizuno, K.; Fujimoto, K.; Shimizu, H.; Inouye, M. *Chem. Eur. J.* 2006, *12*, 824–831. (b) Shimizu, H.; Fujimoto, S.; Maeda, H.; Nanai, Y.; Mizuno, K.; Inouye, Y. *J. Org. Chem.* 2007, *72*, 1530–1533.

(c) Ara, A. M.; Iimori, T.; Nakabayashi, T.; Maeda, H.; Mizuno, K.; Ohta, N. *J. Phys. Chem. B* 2007, *111*, 10687–10696. (d) Tamai, T.; Watanabe, M.; Maeda, H.; Mizuno, K. *J. Polym. Sci. Part A: Polym. Chem.* 2008, *46*, 1470–1475.

[34] *Modern Molecular Photochemistry*, N. J. Turro, Ed.; The Benjamin/Cummings Publishing Co., Inc.: Menlo Park, California, 1978.

[35] Kyushin, S.; Matsuura, T.; Matsumoto, H. *Organometallics*. 2006, *25*, 2761–2765.

[36] Karatsu, T.; Hazuku, R.; Asuke, M.; Nishigaki, A.; Yagai, S.; Suzuri, Y.; Kita, H.; Kitamura, A. *Org. Electronics*. 2007, *8*, 357–366.

In: Environmentally Harmonious Chemistry... ISBN: 978-1-60876-428-0
Editors: M. Anpo and K. Mizuno

Chapter 6

CREATION OF NEW ORGANIC MOLECULES, NEW REACTIONS BASED ON HETEROATOM CHEMISTRY

Akiya Ogawa, Akihiro Nomoto and Motohiro Sonoda

ABSTRACT

In view of green chemistry, the development of organic synthesis with a focus on environmentally safe systems and related technologies have been strongly desired. Along these lines, we have recently revealed a series of eco-friendly synthetic methods based on a clarification of the characteristic features of heteroatom elements. This chapter deals with our recent advances in eco-friendly catalytic reactions, sunlight-induced atom-economical reactions, and environmental harmonious solvents. In chapter 1, our pioneering works on the transition-metal-catalyzed addition reactions of group 15-16 heteroatom compounds are described in detail. Chapter 8-3 deals with eco-friendly oxidation with molecular oxygen based on the development of recyclable catalysts bearing water-familiar ligands. Visible-light-induced radical additions of heteroatom compounds and reductive transformations using rare earth metals and complexes are described in chapters 2 and 4, respectively. Fluorous solvents as environmental harmonious solvents are presented in chapter 5. In addition, some applications of these synthetic methods to material science and life science are introduced in chapter 6.

INTRODUCTION

In organic synthesis directed to green chemistry, the following requirements can be considered: (1) Catalytic use of reagents (*i.e.*, the development of new catalytic systems); (2) Development of atom-economical reactions; (3) High efficiency and high selectivity; (4) The use of safe solvents; and (5) The use of photo-energy, especially visible light (sunlight). Along these lines, we have recently developed a series of new reactions based on a clarification of the characteristic features of heteroatom compounds including the main group elements, transition-metal elements, and rare earth elements. This chapter deals with organic synthesis directed to green chemistry exemplified by our recent research work.

1. TRANSITION-METAL-CATALYZED REACTIONS OF HETEROATOM COMPOUNDS[1]

In principle, addition reactions are atom-economically superior to substitution reactions and elimination reactions, because all of the starting materials are usually involved in the products. In view of green chemistry, therefore, the development of highly selective addition reactions is strongly desired. To attain high selectivity and high efficiency in addition reactions, the use of transition-metal-catalysts is promising. Indeed, a number of examples of transition-metal-catalyzed addition reactions of boron, silicon, and stannane compounds to carbon-carbon unsaturated compounds have so far been developed. In contrast, transition-metal-catalyzed reactions of group 16 heteroatom compounds have been largely unexplored because these group 16 compounds are widely regarded as catalyst poisons. However, we have carried out detailed investigations into the reactivities of transition metal complexes bearing the group 16 functions as ligands toward carbon-carbon unsaturated compounds, and have revealed a series of transition-metal-catalyzed addition reactions of the group 16 heteroatom compounds. Equations 1-3 summarizes the highly selective addition reactions of organosulfur and selenium compounds to alkynes catalyzed by transition metal catalysts.

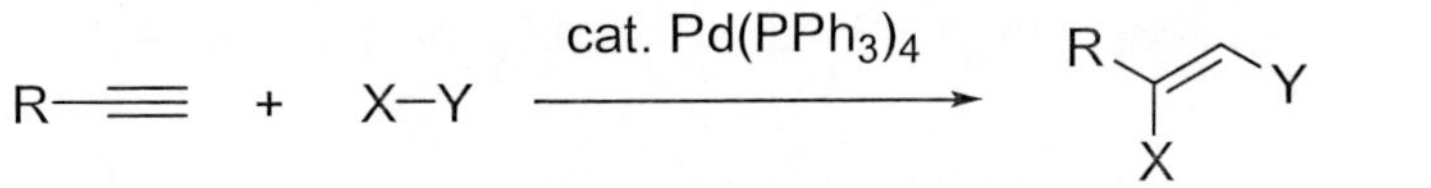

(1)

(X-Y: PhS-SPh, PhSe-SePh, PhSe-SiMe$_3$,
PhSe-GeEt$_3$, PhS-CN)

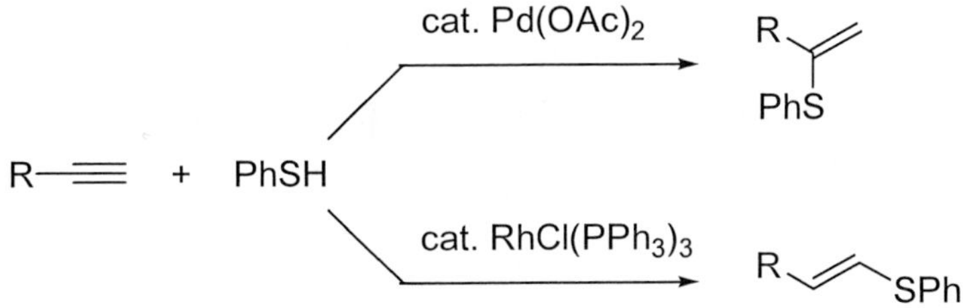

(2)

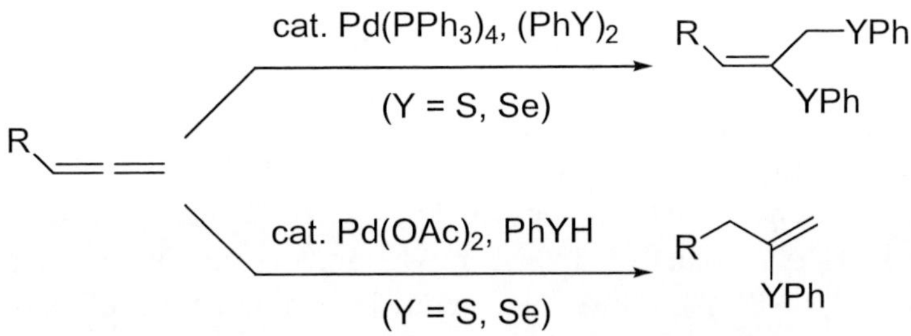

(3)

In the presence of tetrakis(triphenylphosphine)palladium(0), disulfides, diselenides, silyl selenides, germyl selenides, and thiocyanates add to alkynes with excellent stereo- and/or regioselectivities (equation 1). These addition reactions may proceed *via* the oxidative addition of these sulfur and selenium compounds to zero-valent palladium species, followed by the insertion of alkynes into palladium-group 16 atom bonds and the reductive elimination of the corresponding adducts with the regeneration of the catalyst. Hydrothiolation of alkynes with thiols can be controlled with excellent regioselectivity by the selection of the catalysts (equation 2): In the $Pd(OAc)_2$-catalyzed reactions, $Pd(SR)_2$ is formed as a key species *via* the ligand-exchange reaction of the catalyst with thiols, and then the thiopalladation of alkynes and protonation of vinylpalladium species with thiols may take place to give the Markovnikov addition products selectively. On the other hand, *anti*-Markovnikov addition occurs in the $RhCl(PPh_3)_3$-catalyzed reaction, which may proceed *via* the hydrorhodation of alkynes by $HRh(SR)L_n$ (formed in situ by the oxidative addition of thiols (RSH) to the catalyst), followed by the reductive elimination of the products. Similar procedures can be applied successfully to regioselective bisthiolation, bisselenation, hydrothiolation, and hydroselenation of allenes (equation 3).

Under the pressure of carbon monoxide, the site-selective carbonylative addition of sulfur and selenium compounds to alkynes can be catalyzed by several transition metal complexes (Eqs. 4-6). Carbonylative addition of disulfides and diselenides to alkynes and allenes takes place regioselectively to give the corresponding thioesters and selenoesters, respectively (Eqs. 4-5).

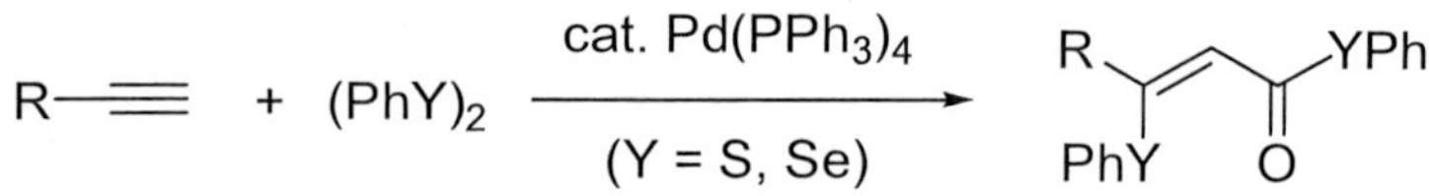

(4)

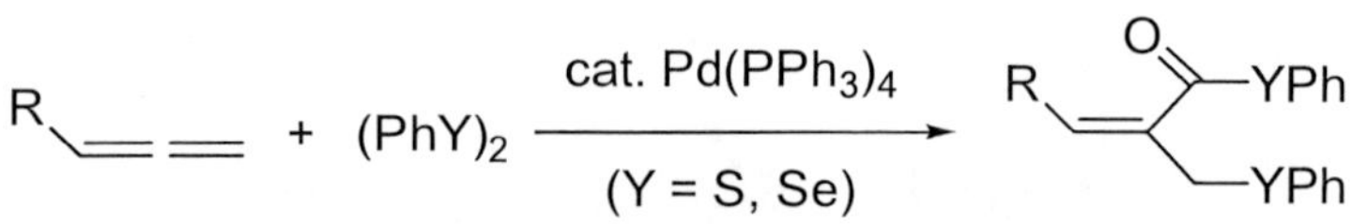

(5)

In the case of thiols, the rhodium-catalyzed reaction provides the corresponding thioformylation products in good yields, whereas the hydrothiocarbonylation of alkynes occurs by use of a platinum(0) catalyst (equation 6). In rhodium-catalyzed thioformylation, the oxidative addition of thiols (RSH) to $RhH(CO)(PPh_3)_3$ and reductive elimination of molecular hydrogen generate RS-$Rh(CO)(PPh_3)_n$ ($n = 2$-3) as a key intermediate. The subsequent thiorhodation of alkynes and CO insertion provide the thioformylation products. In the $Pt(PPh_3)_4$-catalyzed carbonylation, oxidative addition of thiols to Pt(0) and insertion of CO into the S-Pt bond form $HPt(COSR)(PPh_3)_n$ ($n = 2$-3). Acylplatination of alkynes and reductive elimination afford the thioester derivatives. Furthermore, novel thiolative double carbonylation of alkynes with thiols and CO has been developed by employing cobalt carbonyl catalyst. This carbonylation may proceed via the formation of cobalt-alkyne complexes. In the case of allenes as substrates, the hydrothiocarbonylation with thiols and CO takes place in the presence of $Pt(PPh_3)_4$, whereas $RhH(CO)(PPh_3)_3$ or $Pd(OAc)_2$ catalyzes regioselective co-polymerization of allenes and CO preferentially (equation 7).

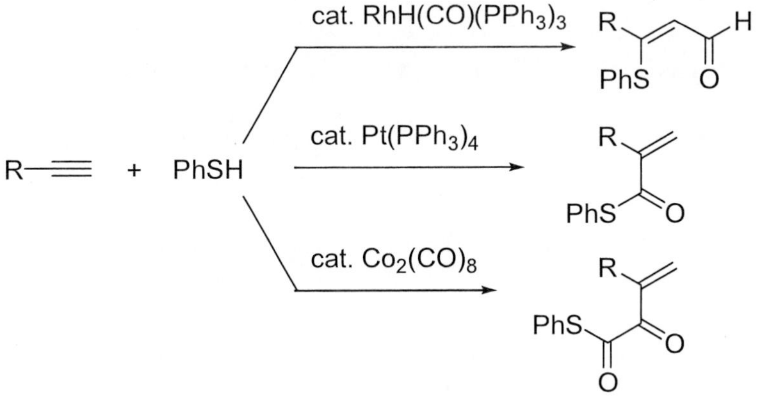

(6)

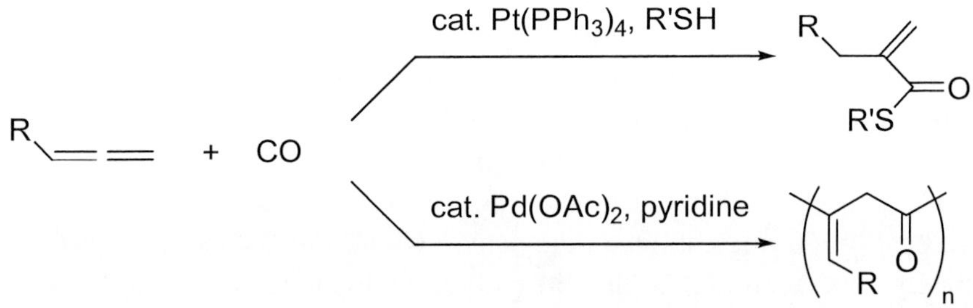

(7)

As mentioned above, the use of transition metal catalysts in the reactions of organosulfur and selenium compounds makes it possible to attain highly selective and atom-economical transformations of these heteroatom compounds. Very recently, these procedures demonstrated in group 16 heteroatom compounds are examined to be applied to group 15 heteroatom compounds such as diphosphines, and the novel hydrophosphination of alkynes has been successfully revealed based on the characteristic features of phosphorus (equation 8). The mechanistic aspects between the differences in the reaction courses between diphosphines and disulfides are now under investigation.

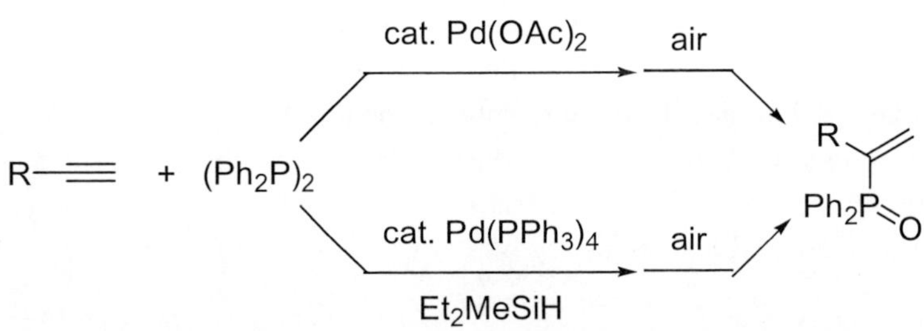

(8)

2. Photoinduced Radical Reactions of Heteroatom Compounds[2]

Not only transition metal catalysts but also photoirradiation can activate heteroatom-heteroatom single bonds. For example, disulfides, diselenides, and ditellurides exhibit their absorption maximum in UV, near-UV, and visible regions, respectively, and therefore, irradiation of the light of these wavelengths causes homolytic cleavage of group 16 heteroatom-heteroatom single bonds, generating the corresponding group 16 heteroatom-centered radicals as labile species. Nevertheless, the photoirradiated radical addition of disulfides, diselenides, and ditellurides to alkenes does not proceed efficiently. This can be explained reasonably by the following kinetic data. The reactivity of the seleno radical toward carbon-carbon unsaturated bonds is relatively lower by a factor of about 10-50, compared with the corresponding thiyl radical. On the other hand, the rate constants for the S_H2 reaction are estimated to be 7.6 x 10^4 $M^{-1}s^{-1}$ for $(PhS)_2$ and 1.2 x 10^7 $M^{-1}s^{-1}$ for $(PhSe)_2$ by using the free radical clock system of 5-hexenyl radical. These kinetic data indicate that $(PhSe)_2$ exhibits an excellent carbon radical capturing ability, compared with $(PhS)_2$, by a factor of approximately 160. Accordingly, the difficulty in realizing the radical addition of disulfides to alkenes may be owing to the lower carbon radical capturing ability of disulfides, whereas the difficulty in realizing the radical addition of diselenides to alkenes may be due to the lower reactivity of seleno radicals toward carbon-carbon double bonds.

Based on these kinetic considerations, we have examined the visible-light-induced radical addition to alkenes by the combination of disulfides and diselenides, and have found novel highly efficient thioselenation of alkenes, in which thio and seleno groups are introduced into the terminal and internal positions of carbon-carbon double bond, respectively, with excellent regioselectivity (Scheme 1). This is because both higher reactivity of thiyl radicals and higher carbon radical capturing ability of diselenides make it possible to attain the efficient radical addition of group 16 heteroatom compounds. Similar procedures can be employed with a wide variety of unsaturated compounds such as alkynes, allenes, conjugated dienes, and isocyanides, giving rise to corresponding thioselenation products regioselectively. The visible-light-induced thioseloenation of enynes proceeds via a radical cyclization process, whereas vinylcyclopropanes undergo thioselenation involving a ring-opening process.

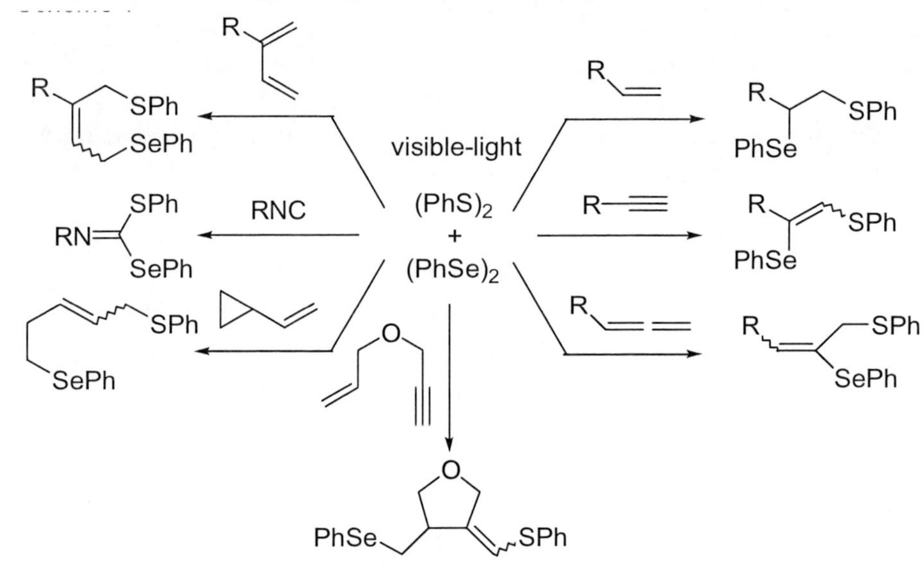

Scheme 1.

More interestingly, diselenides are found to catalyze the radical addition of disulfides to conjugated dienes and isocyanides, which are difficult to be attained with disulfide single systems (equation 9). Similarly, ditellurides can catalyze the photoinduced radical addition of disulfides to allenes.

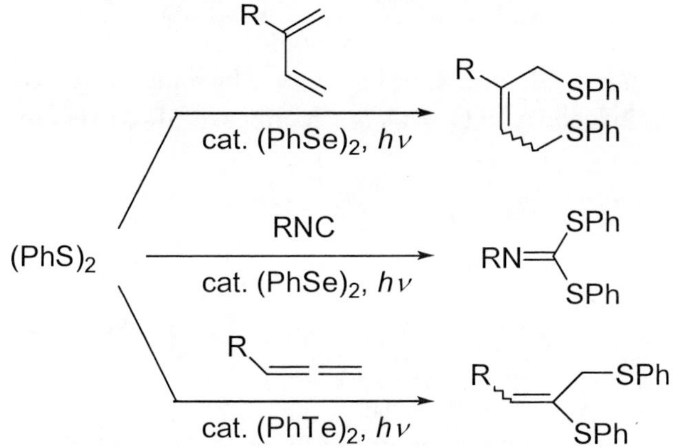

(9)

Success in the thioselenation of enynes prompted us to examine an intermolecular sequential addition to alkynes and alkenes, and the diselenide single system was found to be effective for this purpose because the excellent carbon radical capturing ability of diselenides inhibits the unfavorable polymerization of unsaturated compounds (equation 10). Upon irradiation with visible light, diselenides add to electron-deficient alkyne, electron-rich alkene, and electron-poor alkene, sequentially, *via* the radical cyclization process.

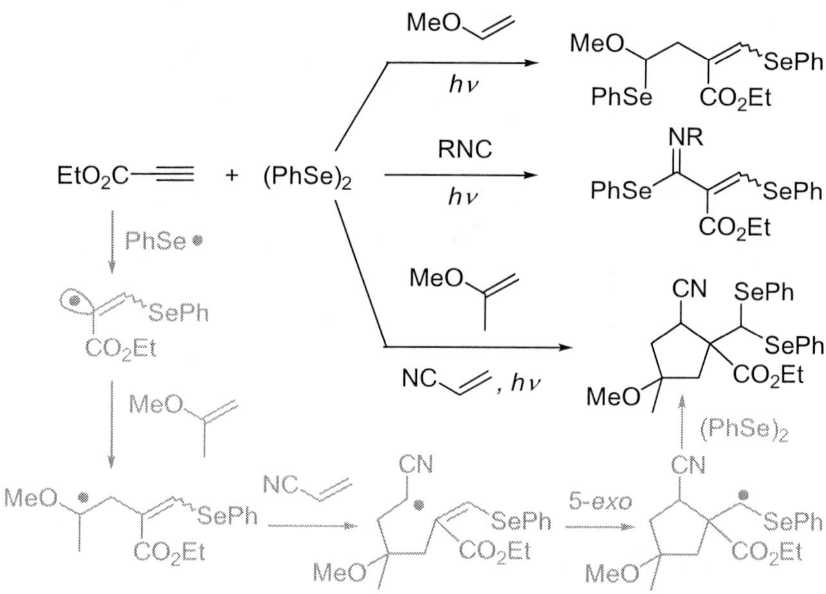

(10)

Diphosphines bearing a group 15 heteroatom-heteroatom bond also add to alkynes upon irradiation with near-UV light, affording the corresponding *vic*-bisphosphinoalkenes in good yields with the preference of *Z*-isomers (equation 11).

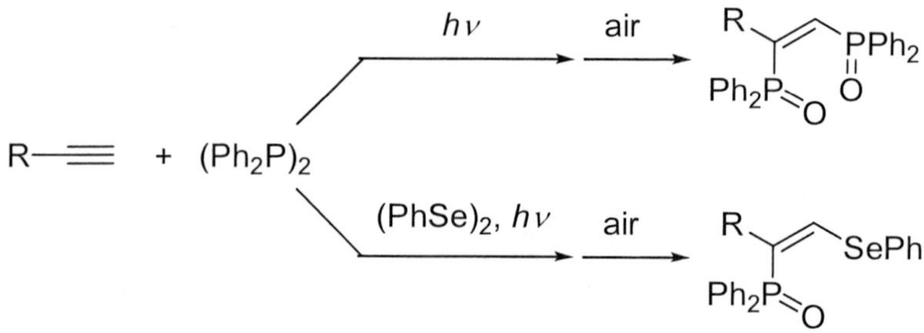

(11)

Furthermore, the novel photoinduced selenophosphination of alkynes takes place by using a diphosiphine-diselenide mixed system.

Most of the addition reactions mentioned above can be induced by the sunlight irradiation, and therefore these reactions are energetically superior to other thermal reactions or radical-initiator induced reactions.

3. Early-Transition-Metal-Catalyzed Oxidation with Molecular Oxygen[3]

Oxidation is an important basic reaction in organic synthesis, and chromium oxides, manganese oxides, and hydrogen or organic peroxides are widely utilized as representative oxidizing reagents. However, these reagents are toxic and/or explosive and stoichiometric or much more amounts of reagents should be required generally. Therefore, oxidations using catalytic amounts of transition metal complexes with molecular oxygen (or air) are extremely desired, and have attracted increasing attention in view of their environmental utility.

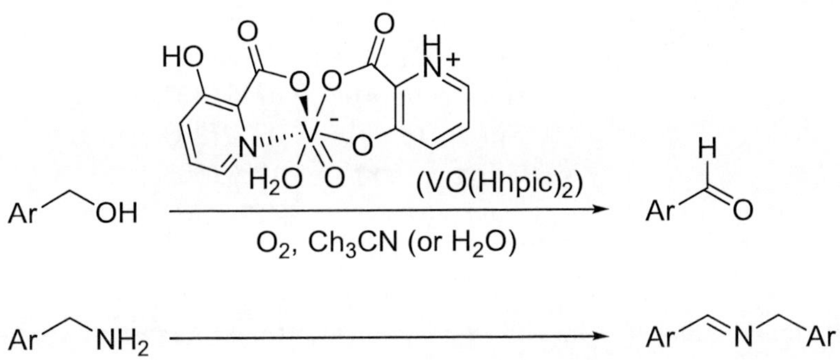

(12)

Vanadium has various oxidation states, and especially, V(V) acts as a useful oxidant. To control the nature of vanadium complexes as single-electron oxidants, the selection of appropriate ligands in vanadium complexes is promising. Recently, we have developed the highly selective oxidation of benzylic alcohols and amines to corresponding aldehydes (or ketones) and imines, where vanadium complexes prepared from $VOSO_4$ and 3-hydroxy-2-picolic acid exhibit excellent catalytic activity under the pressure of molecular oxygen in acetonitrile (equation 12). The catalysts can be recovered easily by removal of the solvent and filtration, and can be reused for the oxidation. More interestingly, the oxidation of alcohols proceeds successfully even in water.

4. Photoinduced Rare Earth Reduction Systems[4]

Recently, we have developed novel visible-light-irradiated reduction systems of low-valent rare earth species. Samarium diiodide (SmI_2 in THF) as a representative divalent species of rare earth halides exhibits its absorption maximum in 565 and 617 nm based on the 4f-5d transition, and therefore irradiation with visible light causes dramatic enhancement of the reducing ability of SmI_2. For example, SmI_2 can reduce organic chlorides smoothly upon visible light irradiation, although the same reductions with SmI_2 do not proceed in the dark. Similarly, other low-valent rare earth species generally undergo photoinduced enhancement of the reducing ability, as exemplified by the reduction of 1-chlorododecane to dodecane (equation 13).

$$ {}^{n}C_{12}H_{25}I \xrightarrow{LnI_2} \xrightarrow{H^+} {}^{n}C_{12}H_{26} \qquad (13) $$

Ln =	Pr, $h\nu$ (dark)	75% (41%)
	Nd, $h\nu$ (dark)	67% (33%)
	Sm, $h\nu$ (dark)	92% (86%)
	Gd, $h\nu$ (dark)	12% (~0%)
	Tb, $h\nu$ (dark)	23% (~0%)
	Yb, $h\nu$ (dark)	60% (41%)

These photoinduced reduction systems of low-valent rare earth species can be applied to the visible-light-induced carboxylation of haloalkanes with carbon dioxide, which affords corresponding carboxylic acids in good yields. The coordination of CO_2 to rare earth species and the photoinduced electron-transfer from the rare earths to the coordinated CO_2 make it possible to attain the novel sunlight-induced fixation of CO_2 (equation 14).

$$ {}^{n}C_{12}H_{25}I + CO_2 \xrightarrow{SmI_2,\ h\nu} \xrightarrow{H^+} {}^{n}C_{12}H_{25}CO_2H \qquad (14) $$

5. Fluorinated Solvent and Synthesis of Fluorous Tags[5]

In view of green chemistry, the selection of solvents is of great importance. In particular, halogenated solvents such as dichloromethane (CH_2Cl_2), which are useful in organic synthesis but are highly toxic, should be replaced by a safe solvent. Recently, benzotrifluoride (BTF: CF_3-Ph) has been employed as an excellent alternative solvent for halogenated and related solvents. In BTF, for example, Aldol condensation, Sakurai-Hosomi reaction, Diels-Alder reaction, Friedel-Crafts reaction, Dess-Martin oxidation, selenoxide *syn*-elimination, and some protection reactions can proceed successfully. BTF is a hybrid solvent and both organic and fluorous substrates and reagents are solved in BTF. Therefore, BTF is also useful for the reactions of fluorous compounds and fluorination reactions of organic substrates. The photoinduced iodoperfluoroalkylation of unsaturated compounds with iodoperfluoroalkanes proceeds in BTF (equation 15). Similarly, the selenoperfluoroalkylation of alkynes is also attained by employing a RfI-$(PhSe)_2$ mixed system.

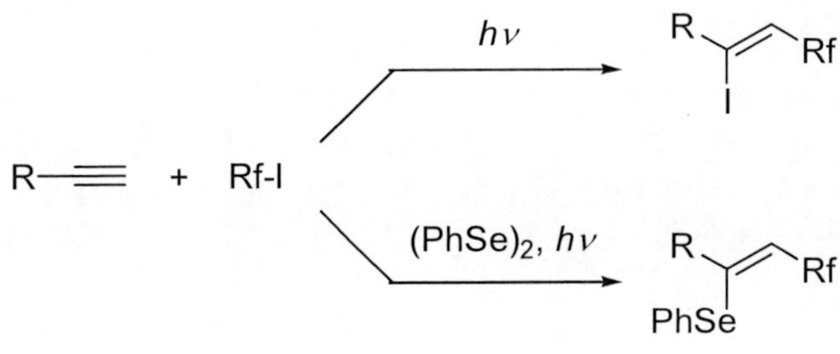

(15)

6. Applications to Material Science[6]

The development of new synthetic methods enables the exploration of a large number of novel functionalized molecules. From the viewpoint of material science, chalcogen atoms are strongly expected to play an important role in the design of new materials. As mentioned already, we have developed a series of novel radical or transition-metal-catalyzed reactions of chalcogen compounds, and therefore, application of these methods to the construction of redox-active molecules functionalized by chalcogen moieties is of great importance because these approaches may lead to the discovery of new functional materials.

According to our established synthetic methods, the reactions of ethynylferrocene with diphenyl disulfide and diphenyl diselenide (or diphenyl ditelluride) have been examined under nitrogen atmosphere upon irradiation with a tungsten lamp (300 W). The desired addition proceeds successfully and affords the corresponding vinylferrocene derivatives in moderate to good yields (equation 16). Comparing the UV-visible spectra of these heteroatom- substituted vinylferrocenes with that of vinylferrocene itself, a remarkable red-shift is observed for the heteroatom-substituted ones. In particular, phenyltelluro-substituted ferrocene shows relatively maximal red-shift. These results indicate that π-conjugation is extended widely through the chalcogen atoms.

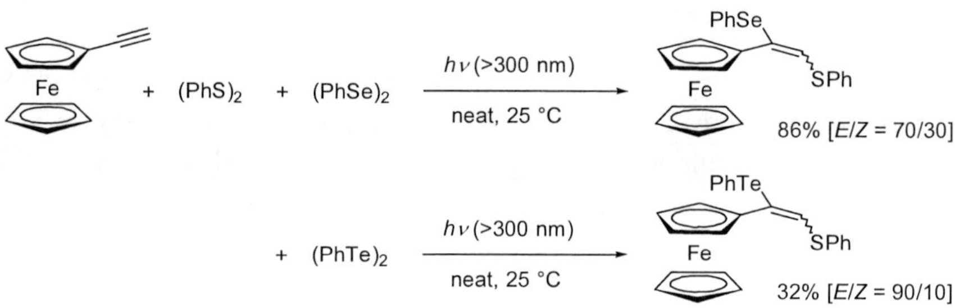

(16)

Sulfur-containing π-conjugated polymers are synthesized by the $RhCl(PPh_3)_3$-catalyzed addition reaction (equation 17). This polymer is soluble in an organic solvent and its ^1H NMR spectrum clearly indicates excellent regio- and stereoselectivities for this reaction.

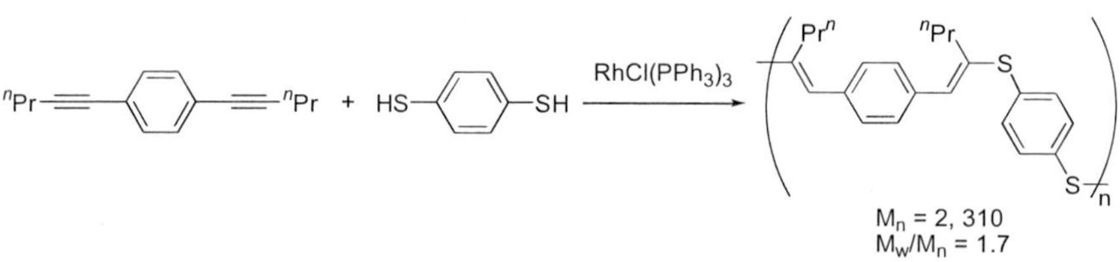

(17)

Not only π-conjugated systems but also σ-conjugated systems (or σ-π conjugated systems) are also attractive because of their characteristic photochemical features. For example, polysilanes can be synthesized by the reduction of dichlorosilanes with the Sm-SmI_2 mixed system. Judging from the luminescence spectrum, the purity of the polysilanes synthesized is high. As exemplified in equation 18, the polymerization of 1,4-bis(dimethylchlorosilyl)-

benzene affords the σ–π conjugated polymeric disilane with a sharp molecular distribution ($M_w/M_n = 1.2$; $M_n = 2,480$).

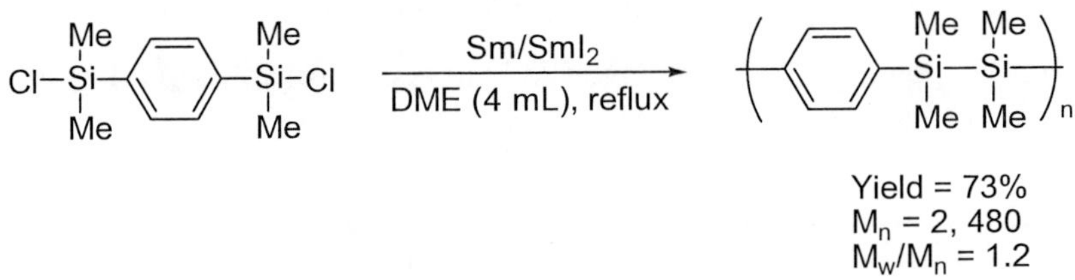

(18)

Conclusion

In organic synthesis, the development of environmental harmonious synthetic reactions will become increasingly important. We believe the use of various heteroatoms in organic synthesis contributes to the efficiency, selectivity, and atom-economy of synthetic reactions, and makes it possible to attain many eco-friendly molecular constructions.

References

[1] For reviews, see: (a) Ogawa, A. In *Handbook of Organopalladium Chemistry for Organic Synthesis*; Negishi, E., Ed.; Wiley: New York, 2002; Chapter VII.6. (b) Ogawa, A. *J. Organomet. Chem.* 2000, *611*, 463. For our recent reports, see: (c) Kamiya, I.; Kawakami, J.; Yano, S.; Nomoto, A.; Ogawa, A. *Organometallics* 2006, *25*, 3562. (d) Kajitani, M.; Kamiya, I.; Nomoto, A.; Kihara, N.; Ogawa, A. *Tetrahedron* 2006, *62*, 6355. (e) Kamiya, I.; Nishinaka, E.; Ogawa, A. *Tetrahedron Lett.* 2005, *46*, 3649. (f) Kamiya, I.; Nishinaka, E.; Ogawa, A. *J. Org. Chem.* 2005, *70*, 696. (g) Kawakami, J.; Takeba, M.; Kamiya, I.; Sonoda, N.; Ogawa, A. *Tetrahedron* 2003, *59*, 6559. (h) Kawakami, J.; Mihara, M.; Kamiya, I.; Takeba, M.; Ogawa, A.; Sonoda, N. *Tetrahedron.* 2003, *59*, 3521.

[2] For reviews, see: (a) Ogawa, A. In *Main Group Metals in Organic Synthesis*; Yamamoto, H., Oshima, K., Eds.; Wiley-VCH: Weinheim, 2004; Vol. 2, pp 813. For our recent reports, see: (b) Kawaguchi, S.; Nagata, S.; Shirai, T.; Tsuchii, K.; Nomoto, A.; Ogawa, A. *Tetrahedron Lett.* 2006, 47, 3919. (c) Tsuchii, K.; Doi, M.; Ogawa, I.; Einaga, Y.; Ogawa, A. *Bull.*

Chem. Soc. Jpn. 2005, *78*, 1534. (d) Tsuchii, K.; Doi, M.; Hirao, T.; Ogawa, A. *Angew. Chem., Int. Ed.* 2003, *42*, 3490. (e) Tsuchii, K.; Tsuboi, Y.; Takahashi, J.; Sonoda, N.; Nomoto, A.; Ogawa, A. *J. Org. Chem.*, 2007, *72*, 415-423.

[3] Ueta, Y.; Nomoto, A.; Yano, S.; Ogawa, A. to be submitted for publication.

[4] For reviews, see: (a) Sumino, Y.; Ogawa, A. *J. Synth. Org. Chem. Jpn.* 2003, *61*, 201. For our recent reports, see: (b) Tomisaka, Y.; Harato, N.; Sato, M.; Nomoto, A.; Ogawa, A.. *Bull. Chem. Soc. Jpn.* 2006, *79*, 1444. (c) Kamiya, I.; Iida, K.; Harato, N.; Li, Z.; Tomisaka, Y.; Ogawa, A. *J. Alloys and Compounds* 2006, *408-412*, 437. (d) Tomisaka, Y.; Tsuchii, K.; Ogawa, A. *J. Alloys and Compounds* 2006, *408-412*, 427. (e) Sumino, Y.; Harato, N.; Tomisaka, Y.; Ogawa, A. *Tetrahedron* 2003, *59*, 10499.

[5] For reviews, see: (a) Nomoto, A.; Ogawa, A. In *Electronic Encyclopedia of Reagents for Organic Synthesis*, ed by Paquette, L.; Fuchs, P.; Crich, D.; Molander, G. John Wiley & Sons, in press. (b) Ogawa, A.; Tuchii, K. In *Electronic Encyclopedia of Reagents for Organic Synthesis*, ed by Paquette, L.; Fuchs, P.; Crich, D.; Molander, G. John Wiley & Sons (2006). For our recent reports, see: (c) Tsuchii, K.; Ueta, Y.; Kamada, N.; Einaga, Y.; Nomoto, A.; Ogawa, A. *Tetrahedron Lett.* 2005, *46*, 7275. (d) Tsuchii, K.; Imura, M.; Kamada, N.; Hirao, T.; Ogawa, A. *J. Org. Chem.* 2004, *69*, 6658. (e) Tsuchii, K.; Ogawa, A. *Tetrahedron Lett.* 2003, *44*, 8777.

[6] (a) Li, Z.; Iida, K.; Tomisaka, Y.; Yoshimura, A.; Hirao, T.; Nomoto, A.; Ogawa, A. *Organometallics*, 2007, *26*, 1212-1216. (b) Ikeda, T.; Nomoto, A.; Fukuzawa, S.; Hirao, T.; Ogawa, A. to be submitted for publication.

In: Environmentally Harmonious Chemistry… ISBN: 978-1-60876-428-0
Editors: M. Anpo and K. Mizuno

Chapter 7

Utilization of Light in Polymer Science

Masamitsu Shirai
Department of Applied Chemistry
Osaka Prefecture University
Sakai, Osaka 599-8531, Japan

1. Introduction

We are presently living in a polymer age and they play a very significant role in everyday living. There are innumerable products based on synthetic polymers, e.g., plastics, fibers, films, elastomers coatings, adhesives, rubber, and cellulose derivatives. A huge amount of synthetic polymers have been produced in various industries and they are used in many areas. One of the advantages of synthetic polymers is their durability. However, in this age of environmental awareness, durability can be considered a curse because polymers, particularly plastics, are a highly visible and detrimental component of society's waste stream. However, there are a variety of ways to solve the problem of polymer waste: making polymers degradable for composing in sanitary landfills; incinerating combustible polymers for energy recovery; recycling; and finding innovative uses for used polymers. Presently, we are using a Voluntary Plastic Container Coding System to aid in the identification of various types of plastics for recycling. The code consists of a series of symbols imprinted or molded on the bottom of plastic containers. Each symbol consists of a number in the center of a three-arrow triangle with identifying letters imprinted below. Recycling methodologies vary from remolding, blending, or the de-polymerization to monomers with subsequent re-polymerization. For example, poly(ethylene terephthalate) bottles are degraded to monomers by hydrolysis, which are then re-polymerized to polyester for fiber in film applications, or reformulated into polyurethane or unsaturated polyesters.

In addition to the recycling of used plastics, it is also important to modify the synthetic processes for polymers from conventional processes to environmentally friendly processes. One example is the preparation method for polycarbonate which is commonly prepared by a polycondensation reaction between sodium salt of bisphenol A and phosgene. However, phosgene is a toxic reagent to handle. As an environmentally friendly process, polycarbonate can also be prepared from bisphenol A and diphenylcarbonate. Preparation of aliphatic polyesters using an enzyme is another example of environmentally friendly processes where catalysts based on heavy metals and hazardous chemicals are not necessary.

Light energy can also be used in polymer synthesis, polymer modification, and polymer degradation. In some cases, a photochemical process is an energy saving process compared to conventional thermal processes. In this chapter, the utilization of light in polymers science is described.

2. Photopolymers

Polymers and oligomers which are photochemically reactive are called "photopolymners". The physical properties of photopolymers can be changed upon exposure to light, especially ultraviolet (UV) light. The photo-induced changes in the physical properties of photopolymers are solubility, hydrophobicity and hydrophilicity, refractivity, thermal and chemical stability, and viscosity as well as others. These changes are widely applied to design functionalized materials, e.g., coatings, adhesives, inks, photoresists, printing plates, imaging materials, and recording materials. There are several types of photo-induced reactions for photopolymers such as photopolymerization, photocrosslinking, and photodegradation which are are commonly used.[1]

Photopolymerization is accomplished using monomers and photoinitiators.[2,3] Vinyl monomers such as acrylate, methacrylate, and styrene are polymerized to obtain polymers using photoradical initiators which can generate radical species upon the irradiation of light. Many types of photoradical initiators sensitive to the desired wavelength of light have been developed. Photopolymerization for the preparation of simple linear polymers is less important because they are usually obtained by the thermal polymerization of corresponding vinyl monomers. However, as will be shown in section **, the photopolymerization technique is very important for the polymerization of multifunctional vinyl monomers in relation to UV curing materials. The photo-induced cationic polymerization of monomers such as vinyl ethers, epoxides, and oxetane derivatives are also quite important. Compounds called "photoacid generator" such as triphenylsulfonium triflate which can generate strong acid on irradiation are used as a photo-initiator. Photochemically generated acid can initiate cationic polymerization. This technique is also important for the polymerization of multifunctional monomers.

Photocrosslinking is an important reaction for the formation of network polymers which are insoluble in solvents. Photopolymers which become insoluble in solvents are commonly used as a negative photoresist in the microelectronics industries. Although dimerization, coupling, and insertion reactions are useful for network formation, photopolymerization is also used.

Photodegradation of the polymer main-chain induces a decrease in the molecular weight while the solubility of the degraded polymer is enhanced. In a special case, photo-induced de-polymerization of the polymers may occur to generate monomers. Some photopolymers bearing functional units in the side-chain can induce the alteration of the hydrophilic or hydrophobic natures of polymers upon irradiation. These types of photopolymers are also used for photoresist applications.[4,5]

3. Reworkable UV Curing Materials[6]

Photocrosslinkable materials are widely used in various applications, e.g., coatings, printing inks, coatings, adhesives, photoresists, and solder masks. The curing process using ultraviolet light has several advantages compared to the thermal curing process. The photo-induced curing process is a rapid and energy saving process. Although photochemically crosslinked materials show excellent physical and/or thermal properties, it is difficult or impossible to thoroughly remove them without damaging the underlying materials. In some cases, however, we need to remove the crosslinked materials from the substrate in order to repair or reuse the substrates. Photocrosslinkable or photo-curable materials which could be removed away from the substrate after use are called reworkable materials. These materials are highly functionalized and environmentally friendly. Here, a design concept for such materials is presented. Furthermore, some examples for the formation of photo-induced networks and the degradation of the networks are described.

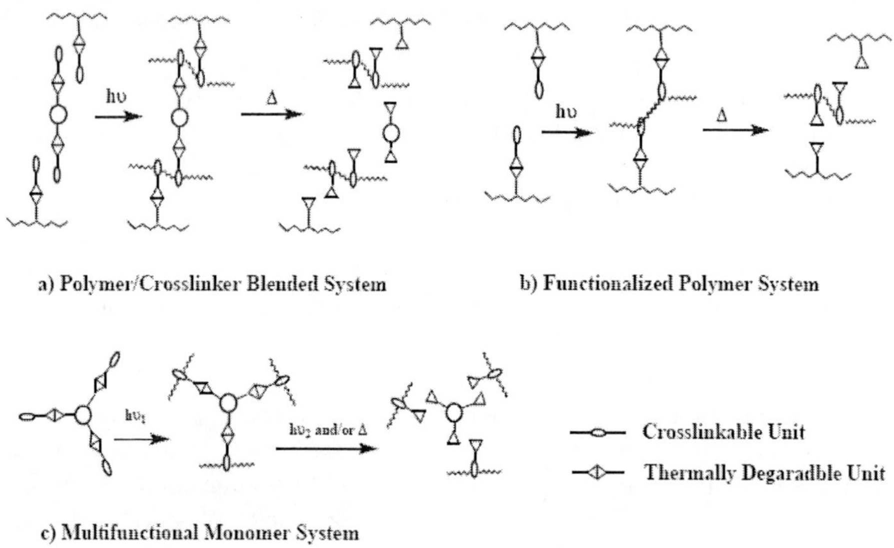

Figure 1. Design concept of a reworkable UV curing system.

Figure 1 shows three types of systems (a)-(c). A blended system of a base polymer and a crosslinker having thermally degradable properties is a convenient system. A thermally degradable crosslinker with reactive units like an epoxy moiety is blended with a photoacid generator (PAG) and a polymer or oligomer having reactive units such as phenol, carboxylic acid, or epoxy units. Photo-induced acid can catalyze the crosslinking reaction between the reactive units of the crosslinker and reactive sites of the polymers. On heating at a given temperature, the scission of the network occurs at thermally degradable units to generate linear polymers which can be dissolved away by the solvents. In the case of system (b), novel polymers containing both crosslinkable units such as epoxy units and thermally degradable units such as tertiary ester units in the side chain are designed. This polymer is used together with PAG. Upon irradiation, the acid generated can induce the crosslinking reactions of epoxy units to form networks. The network is degraded on baking and a linear polymer can be formed. In the case of system (c), multifunctional (meth)acrylate monomers bearing acetal linkages in a molecule are designed. The UV curing can be performed using a conventional photo-radical initiator. The acetal linkages of the cured materials containing PAG are degraded upon irradiation and followed by heating at elevated temperatures.

3.1. Base Polymer/Crosslinker Blended System[7]

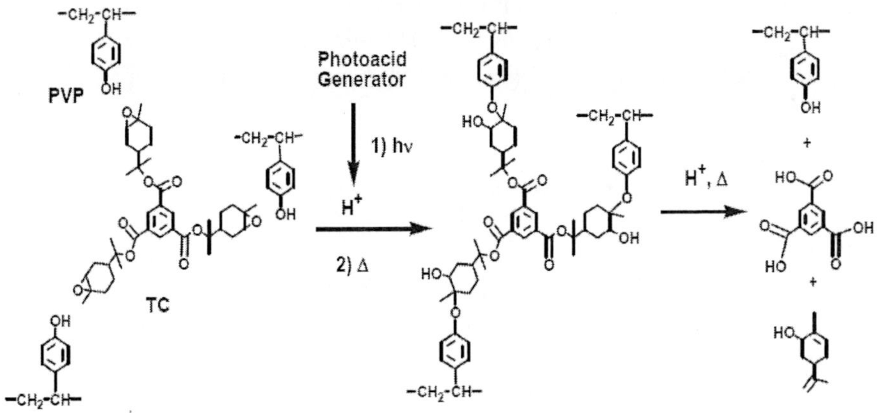

Figure 2. Mechanism for the photoinduced crosslinking and decrosslinking of the PVP/TC blended system.

An example for the base polymer/crosslinker blended system is shown in figure 2. This system consists of polyvinylphenol (PVP), trifunctional crosslinker (TC), and a photoacid generator (PAG). TC is prepared by the reaction of corresponding trifunctional acid chlorides and □-terpineol which is a tertiary alcohol bearing a cycloalkene ring. The olefinic double bond of the cycloalkene ring can easily be converted into an epoxy moiety. The epoxy units work as a photoinduced crosslinking site if we use these compounds in combination with PVP and photoacid generators. Tertiary esters are known to decompose to carboxylic acid and alkene by thermal treatment at temperatures lower than primary and secondary esters.

Multifunctional epoxides are reported to react with phenolic resins to form networks when amines or acids are used as a catalyst. Effective insolubilization of PVP/TC (74 wt% for PVP) blended films containing 5 wt% of a photoacid generator is observed upon UV irradiation with a dose of 60 mJ/cm^2 followed by baking at around 100°C for 10 min. The insolubilization is due to the photoinduced-acid catalyzed crosslinking reaction of the epoxy moiety of TC and phenolic-OH units in the PVP. When the blended films are baked at 120-130°C for 10 min, the crosslinked films become soluble in THF because of the cleavage of the tertiary ester linkages in TC.

The insolubilization efficiency of the blended system is dependent on the PVP/crosslinker ratio. The insoluble fraction of the PVP/TC blend shows a maximum value at 74 wt% of TC. In the blended film, the molar ratio of the epoxy groups against the hydroxyl groups is 0.4. Optimizing the molar fraction of the crosslinkers and baking conditions is important in the present system. It is possible to use polymers containing COOH units instead of PVP because COOH units can react with epoxy crosslinkers in the presence of strong acid.

3.2. Functionalized Polymer System[8-10]

Polymers containing both crosslinkable units such as epoxy units and thermally degradable units such as tertiary ester units in the side chain can be used as a reworkable photopolymer. An example of this type of polymer is shown in Figure 3. This polymer can be obtained by the conventional radical polymerization of a novel monomer, 1-methyl-1-(6-methyl-7-oxabicyclo[4.1.0]hept-3-yl)ethyl methacrylate (MOBH), which has an epoxy moiety and a tertiary ester linkage in a molecule. This polymer is used together with PAG.. Upon UV irradiation, the acid generated can induce the crosslinking reactions of the epoxy units to form networks which are insoluble in solvents and infusible. Photocrosslinking is effectively caused when the PAG which can generate a strong acid is used.

The network is degraded upon baking at given temperatures and a linear polymer can be formed. The network polymer then becomes soluble in solvents.

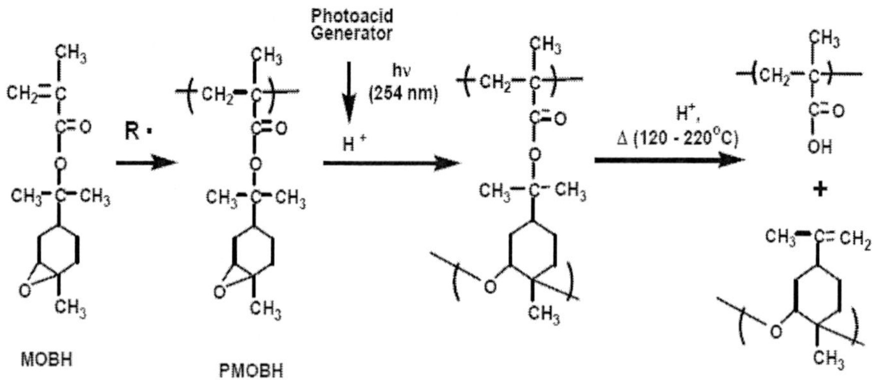

Figure 3. Photocrosslinking and decrosslinking of the reworkable polymer.

The dissolution properties of the crosslinked PMOBH film is shown in figure 4. The PMOBH film containing a photoacid generator, 9-fluorenilideneimino p-toluenesulfonate (FITS), is irradiated at 254 nm with a dose of 60 mJ/cm^2 and the crosslinked film is obtained. The crosslinked film is then baked at 140, 160, and 180 °C and soaked in methanol. An effective dissolution is observed.

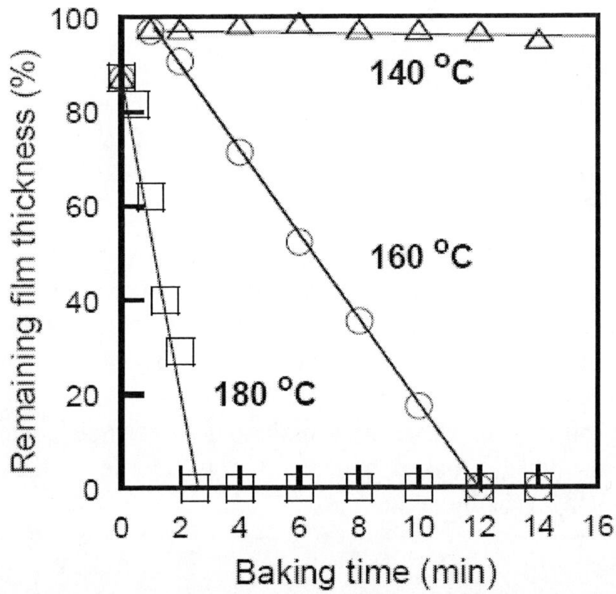

Figure 4. Dissolution property of the crosslinked PMOBH film containing 3.6 mol% FITS.

The copolymers of MOBH and styrenesulfonate derivatives are also interesting materials. They are photocrosslinkable when irradiated with UV light in the presence of PAG. The crosslinked polymer becomes soluble in water when baked at given temperatures. The copolymer of MOBH and cyclohexyl styrenesulfonate (P(MOBH(42)-CHSS)) can be obtained by the conventional radical copolymerization of corresponding monomers. The P(MOBH(42)-CHSS) film containing a PAG becomes insoluble in solvents upon UV irradiation. The crosslinked P(MOBH(42)-CHSS) becomes soluble in water on baking at 120-220 °C. The crosslinked polymers become a water-soluble copolymer of methacrylic acid and styrenesulfonic acid by thermolysis. The reaction scheme is shown in figure 5.

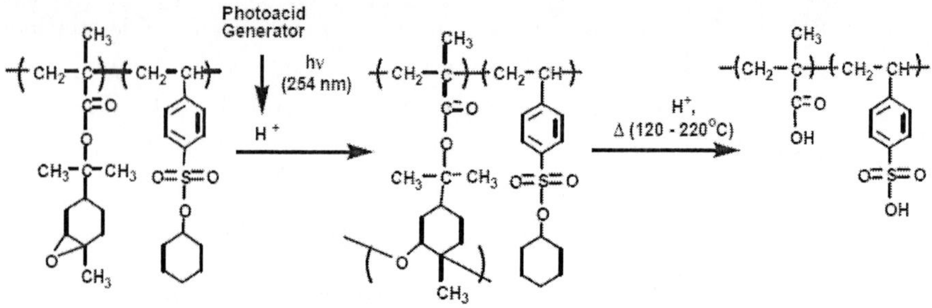

P(MOBH(42)-CHSS)

Figure 5. Mechanism for the photoinduced crosslinking and decrosslinking of a P(MOBH(42)-CHSS) system.

Styrenesulfonates bearing an epoxy moiety are also interesting monomers which can be applied to reworkable UV crosslinking materials. Copolymers based on the monomers shown in figure 6 are photocrosslinkable when used with PAG. The polymer films containing PAG are effectively crosslinked upon UV irradiation. The crosslinked MOPSS(20)-TBMA film becomes soluble in water after baking at 160-220 °C. However, crosslinked copolymers containing TBMA units become insoluble if baked above 220-260 °C due to the formation of carboxylic acid anhydride units.

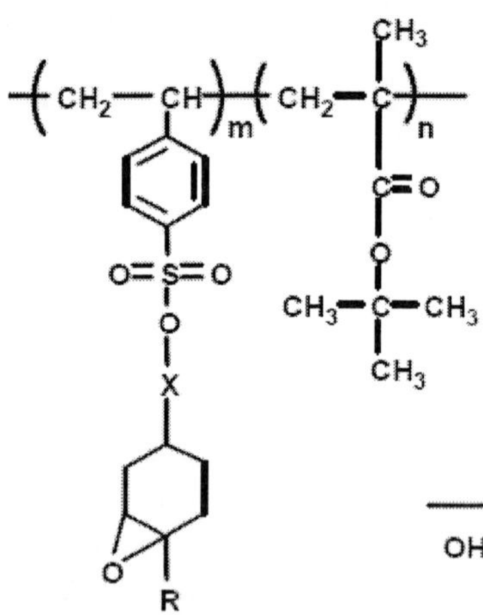

Polymer	X	R
OHMSS(21)-TBMA	$-CH_2CH(CH_3)-$	$-CH_2-$
MOPSS(20)-TBMA	-CH2-	H

Figure 6. Structure of the copolymers.

3.3. Multifunctional (Meth)Acrylate System[11]

Thermally degradable multifunctional methacrylate and acrylate are used as reworkable UV curing materials. The monomers shown in figure 7 are synthesized from corresponding vinyl ether derivatives and hydroxyethyl methacrylate or hydroxyethyl acrylate. The acetal linkage of the monomers is stable up to around 180-240 °C in the absence of acid. Photo-curing of the monomers is performed upon irradiation at 365 nm under N_2 atmosphere using 2,4,6-trimethylbenzoyldiphenylphosphine oxide (TPO) or 2,2-dimethoxy-1,2-diphenylethane-1-on (DMPA) as a photo-radical initiator. Both photo-initiators show almost the same efficiency for photo-curing of the monomers. If irradiation is carried out in air, no curing is observed due to the inhibition effect of oxygen. The curing efficiency as measured by insoluble fraction decreases in the order of: 1b > 1a ~ 2 > 3. The acrylate type monomer is highly curable compared to methacrylate type monomers. Monomer 3 having a bulky structure in the central part of the molecule shows a slow rate. The conversion of acrylate and methacrylate units upon irradiation is measured using the changes of the peak due to the C=C bonds at around 1640 cm^{-1}. The conversion of those monomers at an exposure dose of 200 mJ/cm^2 is easily observed to be about 80 % for 2 and about 60 % for 1b, 1a, and 3.

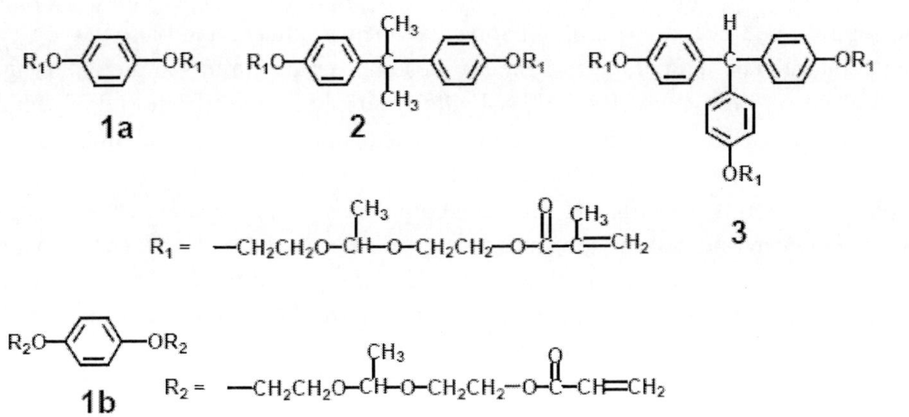

Figure 7. Structures of multifunctional acrylate and methacrylate monomers.

To apply this system to a reworkable UV curing system, samples in film state (1~3 μm) containing 1 wt % of DMPA and 1 wt% of triphenylsulfonium triflate (TPST) was prepared on a silicone wafer and irradiated at 365nm under nitrogen atmosphere to obtain the cured resins. Upon irradiation at 365nm, DMPA decomposes to initiate the polymerization of acrylate or methacrylate units. However, TPST does not decompose because TPST does not absorb light at 365 nm. The cured resins are irradiated at 254 nm (200 mJ/cm^2) followed by baking at various temperatures for 10 min. Upon irradiation at 254 nm, TPST decomposes to generate triflic acid. The acid catalyzes the decomposition reaction of the acetal linkages to generate poly(2-hydroxyethyl methacrylate), acetaldehyde, and diol. The sample is dipped into methanol for 10 min and the thickness of the film remaining on the silicone wafer is measured. About 35-70 % of the cured resins becomes soluble after irradiation at room temperature and more than 80 % of the cured 1a, 2, and 3 becomes soluble after baking at 100-140 °C. The cured resins do not become soluble in methanol if baked at 140 °C without irradiation at 254 nm. A mechanism for UV curing of 2 and photo-thermal degradation of the cured 2 are shown in figure 8. Water, which is necessary to cleave the acetal linkage, is usually supplied from the atmosphere when a thin film is used.

The monomers 1a, 1b, 2, and 3 are also thermally curable and the cured materials are photo-thermally degradable. The monomers containing azobisisobutyronitrile (AIBN) and thermally stable PAG sensitive to 365 nm light are cured upon heating at 80 °C under nitrogen atmosphere. Effective curing occurred and the curing rate for 1b was higher than those for 1a, 2, and 3. Onset temperatures for the decomposition of the cured materials were observed to be 177-236 °C, depending on the structure of monomers. The cured samples decompose and become soluble in solvents when irradiated and baked at a given temperature. The temperature applied to decompose the cured sample is strongly dependent on the type of PAG. If N-hydroxynaphthylimide triflate (NITf) which can generate triflic acid on irradiation is used, most of the acetal linkages are cleaved upon exposure at room temperature. On the other hand, baking at 120 °C after irradiation is necessary to induce a complete cleavage of the acetal linkages if 3-isopropyl-thioxanthone oxime p-toluenesulfonate (ITXTS), which photochemically generates p-toluenesulfonic acid, is used as the PAG. Thus, the temperature for the photo-induced degradation of the cured materials can be determined by choosing the PAG.

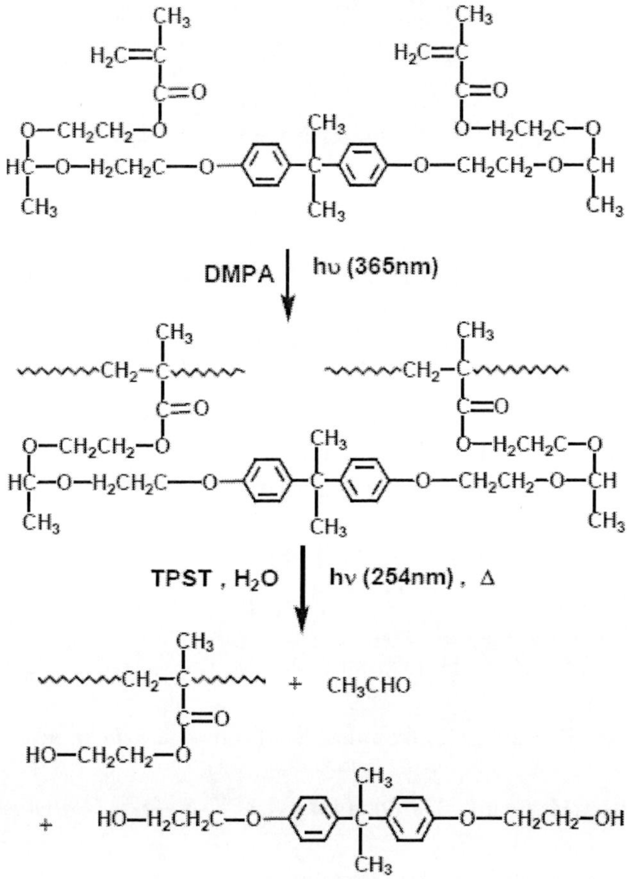

Figure 8. Mechanism for the UV curing and thermal degradation of 2.

5. Conclusion

This chapter dealt with one example of technologically useful light-associated reactions of polymeric and monomeric materials. UV curing materials are widely used as coating materials, printing inks, adhesives, photoresist, and solder masks, because the cured materials show excellent physical and/or thermal properties. However, it is very difficult or impossible to remove the cured materials or network materials from the substrate without damaging the under layer. In some cases, it is necessary to remove the cured materials from the substrate. Here, UV curing materials that are removable from the substrate after use are described. These are highly functionalized and environmentally friendly materials.

References

[1] A. Ravve, Light-Associated Reactions of Synthetic Polymers, Springer, New York, 2006.
[2] J. P. Fouassier, J. F. Rabek, Radiation Curing in Polymer Science and Technology, *Elsevier Applied Science,* New York, 1993.
[3] J. P. Fouassier, Photochemistry and UV Curing Trends, Research Sign Post, Kerala, India, 2006.
[4] L. F. Thompson, C. G. Willson, M. J. Bowden, Introduction to Microlithography, American Chemical Society, Washington, DC., 1994.
[5] R. Arshady, Desk Reference of Functional Polymers, American Chemical Society, Washington, DC., 1996.
[6] M. Shirai, *Prog. Org. Coatings,* 58, 158 (2007).
[7] H. Okamura, K. Shin, M. TSunooka, M. Shirai, *J. Polym. Sci.:Part A: Polym. Chem.,* 42, 3685 (2004).
[8] M.Shirai, S. Morishita, H. Okamura, M. Tsunooka, *Chem. Mater.,* 14, 334 (2002).
[9] M. Shirai, A. Kawaue, H. Okamura, M. Tsunooka, *Chem. Mater.,* 15, 4075 (2003).
[10] M. Shirai, A. Kawaue, H. Okamura, M. Tsunooka, *Polymer*, 45, 7519 (2004).
[11] M. SHirai, K. Mitsukura, H. Okamura, M. Miyasaka, *J. Photopolym. Sci. Technol.,* 18, 199(2005).

In: Environmentally Harmonious Chemistry... ISBN: 978-1-60876-428-0
Editors: M. Anpo and K. Mizuno

Chapter 8

POLYMER-BASED NANOMATERIALS FOR DRUG DELIVERY

Kenji Kono

ABSTRACT

The accurate delivery of drugs and bioactive molecules to a target site of the body is one of the keys of medical technology for the next generation. An approach from polymer chemistry is an efficient method in the production of functional nanomaterials for drug delivery. In this chapter, the design, preparation, and properties of polymer-based nanomaterials with rational functions, such as polymer-drug conjugates, dendrimers, polymer micelles, and polymer-modified liposomes, are outlined.

INTRODUCTION

Therapeutics using chemical drugs are increasing their importance for the treatment of diseases such as cancer. When drug molecules are injected into the body or orally administered, these molecules are usually diffused throughout the whole body. Drug molecules, which reach the disease sites, exhibit therapeutic effect, whereas those which reach normal, healthy sites, can induce undesirable side effects. To obtain the maximum therapeutic effects and minimum side effects, drug molecules must be precisely delivered to the site of the disease.

There may be several strategies for such a specific attack on disease sites using drug molecules. The use of drugs that can specifically reach the disease sites is ideally the simplest way. In fact, many researchers are attempting to realize this idea using antibody-based drugs. Drugs that exhibit therapeutic effects only at the target cells of the disease tissues are also an attractive approach for target specific treatment. However, most drugs lack such specificity to disease sites in the body. Therefore, appropriate carriers are necessary for these drugs to be guided to the target sites. Indeed, carriers must possess a size in the order of nanometers because they need to circulate in the blood stream, permeate through blood vessel walls and diffuse in extracellular spaces before arriving at the target site. In addition, carriers should have rational functions which support the arrival of drug molecules to the site of their action. This chapter will describe a rational design for polymer-based materials of nanometer dimensions for accurate delivery of drug/biologically active molecules in the body and/or inside the cells.

Strategy of Drug Targeting

Approaches to the accurate delivery of drugs to the target sites, i.e. drug targeting, are conceptually divided into two categories: passive targeting and active targeting. Passive targeting achieves the accumulation of carrier-drug complexes at the target sites using the interactions between the carriers and body. It is known that physicochemical properties, such as the chemical structure, charge, and size of the carriers affect their distribution in the body. For example, macromolecules with a molecular weight of less than about 30,000 are filtered from the kidney and, hence, removed from blood circulation, whereas those with higher molecular weights are not filtered from the kidney and exhibit longer circulation time [1,2].

Recently, it was found that macromolecules and particles can effectively accumulate at tumor tissues. Tumor tissues are rich with vessels, the walls of which exhibit higher permeability against large molecules and particles than those of normal tissues [3]. This phenomenon is called the “enhanced permeation and retention (EPR) effect”. Various types of polymers and particles with long circulating properties have been shown to accumulate efficiently at tumor sites and, hence, the EPR effect is being considered as a basis for the design of carrier systems with target specific properties [4].

On the other hand, drug carriers with target specificity have been prepared using biologically specific interactions such as antigen-antibody interactions and ligand-receptor interactions. Tumor-associated antigens and receptors, which are specifically expressed or over-expressed by tumor cells, are useful in specifying the targets. Carriers with these tumor-specific antibodies and ligands are, thus, able to recognize their targets and deliver drugs. Because these systems utilize biological recognition for their ability to zoom into the target sites, this approach is called active targeting.

An alternative approach to active targeting is to use carriers with stimuli-responsive properties. Physical stimuli, such as temperature (heat), ultrasound, light, magnetic fields, and electric fields, are given to the carriers. This approach is very attractive because distribution of the carriers and/or drug release from the carriers inside the body can be controlled from outside the body. Environment-sensitive properties are also used for target specific drug delivery. pH-Sensitive carriers can deliver drugs to the target regions with a weakly acidic environment, such as tumor and inflammation sites. Carriers with environment-sensitive properties also have advantages for the intracellular delivery of drugs, because they can release drugs in intracellular spaces, such as endosomes and lysosomes with weakly acidic environments, and cytosol with a reductive-environment. The rational design of carriers based on a combination of passive targeting and active targeting is essential in increasing the accuracy of drug delivery.

Polymer-Drug Conjugates

The conjugation of drugs to polymers can yield a long-circulating property which enhances accumulation of the conjugates at the target tumor sites through the EPR effect. The incorporation of targeting ligands to the conjugates increases the precision of drug delivery through the synergy of the EPR effect and ligand-receptor interaction. The ligand-receptor interaction may also enhance their internalization into the target cells.

Various kinds of naturally-occurring and synthetic polymers have been used for conjugation with drugs [5]. Typical examples are vinyl polymers such as N-(2-hydroxypropyl)methacrylamide copolymers, styrene-maleic acid copolymers, and divinylether-maleic acid copolymers; poly(amino acid)s such as poly(L-lysine) and poly(L-aspartic acid); polysaccharides such as dextran, inulin, and chitosan; and proteins such as albumin.

From the viewpoint of polymer backbone topology, hyper-branched polymers and dendrimers can be considered of importance since these polymers with branched structures show unique features which conventional linear polymers do not possess (figure 1) [6]. Especially, dendrimers, which are a new type of synthetic polymer with a regularly branched backbone structure, are highly attractive due to their very narrow molecular weight distribution and globular shape. While widely-used poly(amidoamine) (PAMAM) and poly(propylenimine) dendrimers are known to be toxic to cells [7], biodegradable and biocompatible dendrimers, such as polypeptide dendrimers and polyester dendrimers, are being actively developed [8,9]. By utilizing many terminal functional groups, large amounts of drug molecules can be conjugated to the dendrimers [10,11].

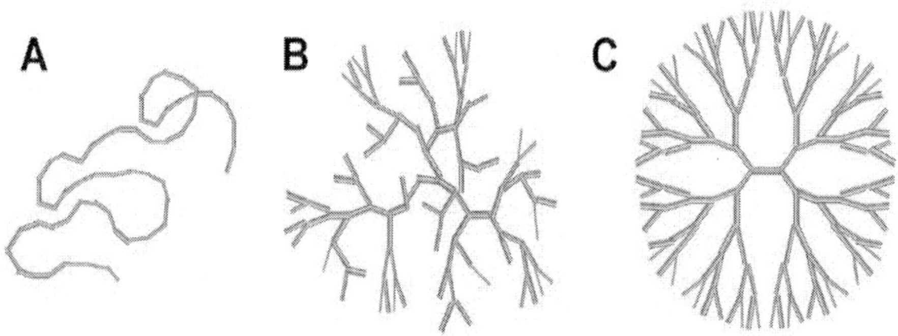

Figure 1. Schematic illustrations of linear polymer (A), hyperbranch polymer (B), and dendrimer (C).

Another significant advantage of dendrimers is the capability of their interiors to encapsulate guest molecules. The efficiency of drug encapsulation by dendrimers depends on the strength of their interactions. PAMAM dendrimers are shown to encapsulate hydrophobic molecules with negatively charged groups, such as methotrexate and rose bengal, whereas poly(glycerol) dendrimers retain a hydrophobic anticancer drug, taxol [12-14]. In addition, metal particles are also shown to be held in dendrimers. Especially, dendrimers encapsulating gold nano-particles are attractive because these composites might generate heat upon visible or infrared light irradiation. Such a property of the composite can be used for photo-thermal therapy [15].

Modification of the terminal reactive groups is a common strategy for the functionalization of dendrimers. Polymer chains are readily attached to all chain ends of dendrimers. Moreover, attachments of poly(ethylene glycol) (PEG) chains provide biocompatible properties to dendrimers. Because the dendrimer moiety of the PEG-modified dendrimers can retain small molecules, they can be used as drug carriers. In addition, the shell of PEG chains may prohibit interaction of the dendrimer moiety with proteins and cells in the body and may accumulate at a target site through the EPR effect.

PEG-modified PAMAM dendrimers were shown to be capable of encapsulating various small molecules including anticancer drugs doxorubicin and methotrexate [12] and photosensitizing agents rose bengal and protoporphyrin IX (figure 2) [13]. The encapsulation efficiency of the guest molecules by the PEG-modified dendrimers depends on the strength of their interactions. PEG-modified PAMAM dendrimers exhibit strong affinity toward molecules with hydrophobic backbone and negatively charged groups. For example, the PEG-modified PAMAM dendrimers have higher encapsulation ability toward methotrexate than doxorubicin [12]. However, PEG-modified PAMAM dendrimers release these guest molecules quite easily under biological conditions. This fact indicates that the encapsulation ability of PEG-modified PAMAM dendrimer is not high enough and needs to be improved for their practical use such as in drug delivery systems [12].

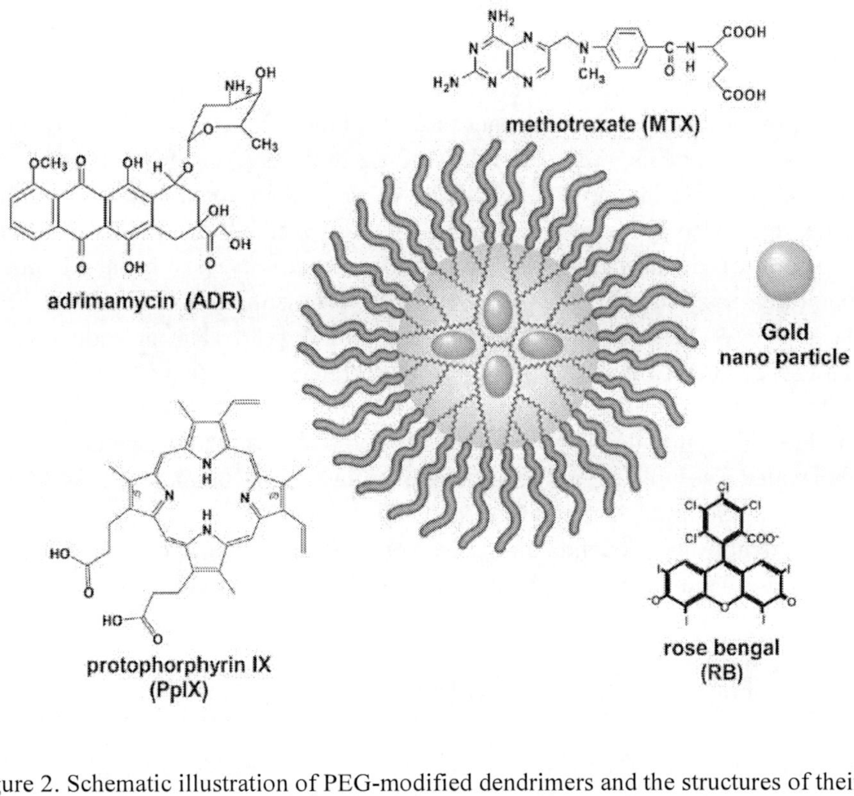

Figure 2. Schematic illustration of PEG-modified dendrimers and the structures of their guest molecules.

The encapsulation ability of PEG-modified dendrimers is improved by incorporating hydrophobic amino acid residues, such as glutamic acid γ-benzyl ester, into the periphery of the dendrimer moiety. Densely packed amino acid residues form a shell in the periphery of the dendrimer, which might suppress diffusion of the encapsulated molecules into the outer phase [16]. Cysteine residues are also incorporated into the dendrimer terminal chains to provide an environment-sensitive property to the dendrimers. The cysteine residues attached to the chain terminals of the dendrimer form a network through disulfide linkages of their side chains in an oxidative environment. However, these linkages are cleaved in a reductive environment, resulting in the disappearance of the cysteine network. Such formation and disappearance of the cysteine network in the dendrimer periphery may control retention and release of the guest molecules [17]. These functions of dendrimers may enhance the usefulness of dendrimers as drug delivery vehicles.

Stimuli-sensitive properties are of importance for carriers to achieve the accurate delivery of drugs. Many efforts have been made to provide the optimal pH, temperature, light and other factors involved. Because the interior of PAMAM dendrimers contains many tertiary amines, these dendrimers show pH-sensitive properties. If such dendrimer encapsulating drugs are taken up by a cell through the endocytosis, they will release drug molecules in the endosomes and lysosomes, which contain weakly acidic environments. Temperature-sensitive properties may also be very useful from a more practical viewpoint so that the development of dendrimers with temperature sensitivity has been attempted. Temperature-sensitive polymers, such as poly(N-isopropylacrylamide), were attached to the surface of dendrimer molecules [18]. The introduction of alkylamide groups to the periphery of dendrimers was also shown to give temperature-sensitive properties to the dendrimers (figure 3A) [19-21]. These alkylamide-terminated dendrimers are soluble in water at a low temperature, but became water-insoluble at a specific temperature. This temperature is termed lower critical solution temperature (LCST). The LCST of the alkylamide-terminated dendrimers depends on the hydrophobicity of the terminal alkylamide groups. These dendrimers can encapsulate small guest molecules and still retain their temperature-sensitive properties. By using these properties of the dendrimers, recovery of the guest molecules can be achieved by increasing the temperature above the LCST (figure 3B) [22].

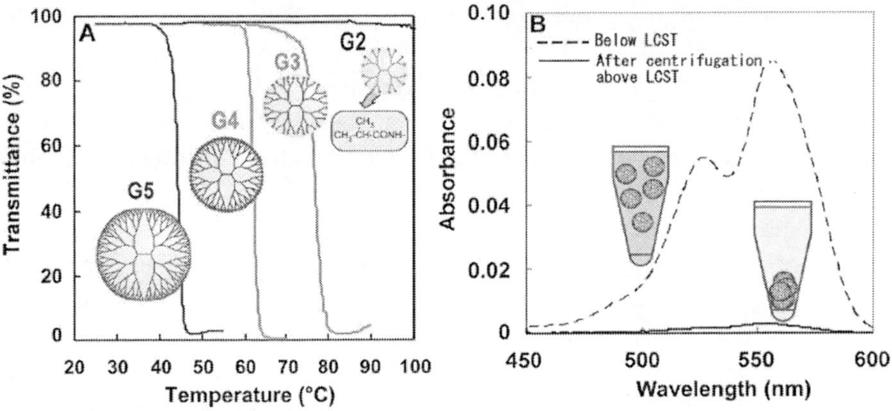

Figure 3. Temperature-sensitive properties of isobutyramide-terminated PAMAM dendrimers. (A) Transmittance of aqueous solutions of isobutyramide-terminated PAMAM G2-G5 dendrimers as a function of temperature. (B) Absorption spectra of an aqueous solution of isobutyramide-terminated G5dendrimer containing rose bengal below LCST or above LCST with centrifugation.

Polymer Micelles

Amphiphilic polymers, which consist of hydrophobic segments and hydrophilic segments, form a micellar structure in aqueous solutions. Typical examples of amphiphilic polymers are block copolymers composed of a hydrophilic block and hydrophobic block. Like low-molecular-weight detergents, amphiphilic polymers form micelles only when their concentration is higher than a certain value, which is termed critical micellar concentration (CMC). It is known that the micelles of detergents have a dynamic property: association and dissociation of detergent molecules take place in the milliseconds in the micelles. However, amphiphilic block copolymers might form micelles with a more static character through stronger interactions between the hydrophobic blocks of the copolymers.

For drug delivery purposes, poly(ethylene glycol) (PEG) is widely used as the hydrophilic block of amphiphilic block copolymers. It is known that PEG does not have immunogenicity. In addition, highly hydrated and flexible PEG chains forming a corona of the polymer micelles suppress blood components and cells and improve their colloidal stability in the blood. On the other hand, various types of polymers with a hydrophobic nature, such as poly(propylene oxide), poly(glutamic acid) esters, poly(caprolactone) and poly(lactic acid), are used as a hydrophobic part of amphiphilic block copolymers. These polymer micelles have a core with a hydrophobic character and, hence, can retain hydrophobic drugs, such as paclitaxel and doxorubicin, in the core through hydrophobic interactions. Polymer micelles with a diameter of around several tens nm are shown to accumulate in tumor tissues through the EPR effect after injection into the blood stream. Therefore, polymer micelles containing antitumor drugs can efficiently suppress tumor growth. Moreover, some targeting ligands, such as folate and antibodies, can be conjugated to polymer micelles to improve their target-specificity.

Another strategy to enhance target specific delivery is to provide stimuli-sensitive properties to polymer micelles. Thermosensitive polymers are often incorporated to polymer micelles. Poly(N-isopropylacrylamide) (PNIPAM), which is a typical thermosensitive polymer, exhibits LCST at 32 °C: PNIPAM is highly soluble in water at low temperature but it becomes water-insoluble and hydrophobic above 32°C. Therefore, copolymers of the PNIPAM block and poly(styrene) block form polymer micelles with temperature-responsive properties. For example, these polymer micelles release drug molecules above the LCST. Also, these micelles enhance adsorption onto cells above the LCST. Hydrophocbicity changes in their surface induced by dehydration of the PNIPAM chains can control such drug release and cellular affinity.

POLYMER-MODIFIED LIPOSOMES

Liposomes are vesicles consisting of phospholipid bilayer membranes. They can retain hydrophilic molecules and hydrophobic molecules in their aqueous interior and in their bilayer membrane, respectively. In addition, liposomes possess various advantages as drug carriers such as their excellent biocompatibility and biodegradability. Therefore, their application to drug delivery has been actively studied.

When injected into the body, liposomes are quickly taken up by a reticuloendotherial system (RES), such as the liver and spleen, and are removed from blood circulation. This situation decreases the chance of liposomes reaching the target sites in the body. However, surface modification of liposomes with poly(ethylene glycol) (PEG) may protect them from capture by RES and increase their circulation time, resulting in more efficient accumulation in the target tumor sites through the EPR effect. Therefore, PEG-modified liposome encapsulating anticancer drugs are now being used for cancer chemotherapy.

To elevate the usefulness of liposomes as drug delivery systems, their functionalization is essential. Therefore, many efforts have been made to provide various functions to liposomes. For example, the incorporation of antibodies have been attempted to increase target specificity. Synergy of the EPR effect and antibody-based affinity to the target cells can greatly improve accumulation of liposomes at the target sites [23]. Some other molecules, such as transferrin and folic acid, are also used as liposomal ligands to increase target specificity of liposomes since their receptor molecules are over-expressed on many types of tumor cells [24,25].

Another important function of liposomes is their stimuli-sensitive property. While various types of stimuli-liposomes have been developed, temperature-sensitive liposomes and pH-sensitive liposomes have been most intensively investigated. These liposomes are prepared by surface modification of the liposomes with temperature-sensitive or pH-sensitive polymers.

Temperature-sensitive liposomes are those that can retain drug molecules at physiological temperature, but release them a few degrees above that temperature. These liposomes have originally been prepared using phospholipids whose membrane undergoes gel-to-liquid-crystalline transition at around 42 ^{o}C. However, modification with temperature-sensitive polymers has been shown to be an alternative efficient method to produce temperature-sensitive liposomes (figure 4) [26]. Although poly(NIPAM) exhibits LCST at 32 ^{o}C, its LCST can be adjusted at around 40 ^{o}C by copolymerization with hydrophilic comonomers. For fixation of temperature-sensitive polymers, hydrophobic moieties such as didodecylamide groups were incorporated at random positions or the terminals of polymer chains [27,28]. These hydrophobic groups are inserted into the liposome membranes and act as anchors. Temperature-sensitive polymers change their property between hydrophilic and hydrophobic at LCST. Therefore, when fixed onto a liposome surface, highly hydrated polymer chains increase hydration of the liposome surface and stabilize the liposomes below LCST. However, above LCST, the dehydrated polymer chains cause destabilization of the liposome membrane, resulting in content release and/or membrane fusion. Because drug release and fusion of these liposomes can be controlled by local heating at the target sites, temperature-sensitive polymer-modified liposomes are useful for target site-selective drug delivery and intracellular drug delivery (figure 3) [28-30].

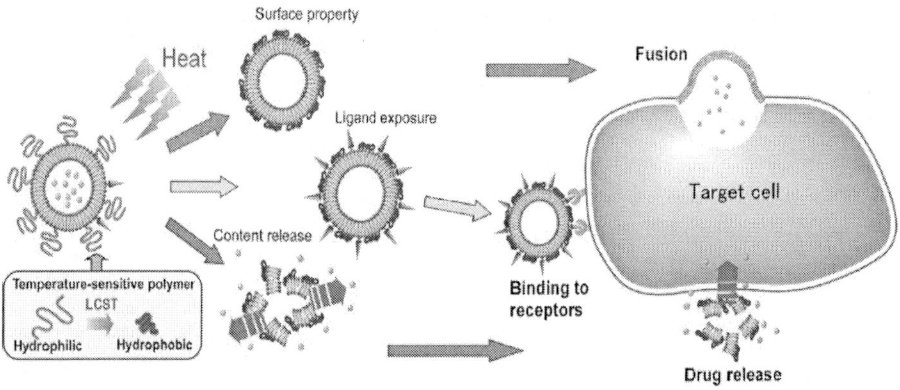

Figure 4. Design of various types of temperature-sensitive liposomes using temperature-sensitive polymers.

pH-sensitive liposomes are liposomes that destabilize under weakly acidic conditions. In general, liposomes are mainly taken up by cells through endocytosis and trapped inside of an endosome which contains a weakly acidic environment. When taken up by a cell, pH-sensitive liposomes become destabilized and fuse with the endosome, resulting in the transfer of the contents into the cytosol. Therefore, pH-sensitive liposomes are used for the cytoplasmic delivery of membrane-impermeable molecules, such as proteins, oligonucleotides and genes [31]. pH-sensitive liposomes are obtained by the surface modification of liposomes with pH-sensitive polymers. For example, poly(ethylacrylic acid) is attached onto the liposome surface. This polymer is negatively charged at pH 7.4 and hence does not interact with the liposome membrane. However, the protonation of carboxylate groups increases interactions between the polymer chains and liposome membrane, which causes intensive destabilization below pH 6 [32].

Other types of pH-sensitive polymers are prepared by the succinylation of poly(glycidol), which has a backbone with a PEG-like structure [33]. Attachment of this polymer can provide a pH-sensitive fusion ability to stable liposomes. When succinylated poly(glycidol) (SucPG)-modified liposomes encapsulating fluorescent dye molecules were added to tumor cells and incubated for several hours, they displayed diffuse fluorescence under observation with fluorescence microscopy. Indeed, when pH-insensitive liposomes containing the same fluorescent molecules were used for cellular treatment, only vesicular fluorescence was observed, suggesting that liposomal contents were still inside the endosome. These facts demonstrate the importance of pH-sensitive properties for cytoplasmic delivery of membrane-impermeable molecules [34].

The pH-Sensitive fusogenic property of SucPG-modified liposomes are used to prepare efficient gene carriers [35]. Gene carriers must pass through various barriers inside a cell before reaching the nucleus, where transcription of a gene takes place. Transfer of the genes from endosome to cytosol is one of the most important processes which strongly affect gene expression efficiency. If this process does not proceed efficiently, then gene-carrier complexes are likely to be degraded in the lysosome. Therefore, promotion of the gene transfer from endosome to cytosol is a key to achieving the efficient transfection of cells. This process is called “endosomal escape”. Fusion between the liposome and endosome is effective in enhancing the endosomal escape of genes. Complexes of cationic lipids and genes are termed lipoplexes. Lipoplexes have highly condensed DNA and a positively-charged surface. They interact strongly with the cell surface via electrostatic force and are efficiently taken up by a cell through endocytosis. Lipoplexes are widely used for gene transfection of cells, although their activity is insufficient. When pH-sensitive liposomes are added to lipoplexes, they form hybrid complexes which generate fusion activity at weakly acidic pH. Indeed, these hybrid complexes achieve efficient transfection toward various types of cells because fusion of the complexes and endosome promotes the transfer of genes into the cytosol (figure 5) [36-38].

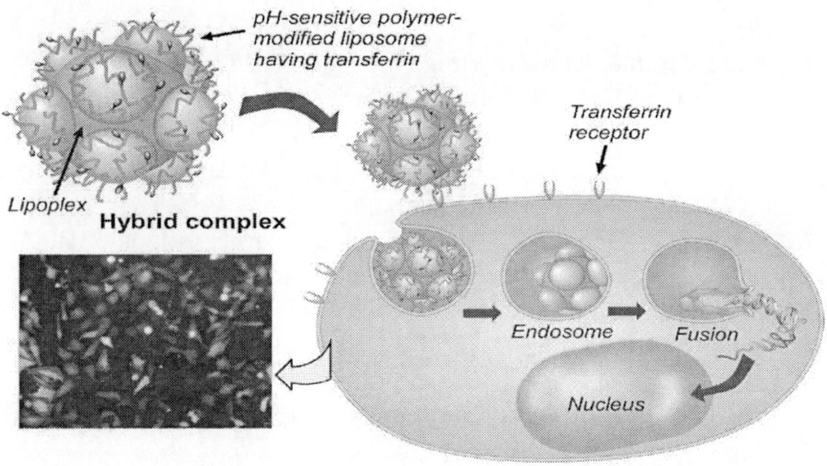

Figure 5. Schematic illustration of gene delivery by a hybrid complex of pH-sensitive fusogenic liposome and lipoplex. Fluorescence micrograph of HeLa cells treated with a hybrid complex containing a green fluorescent protein gene is also shown.

Conclusion

Nanomaterials are expected to induce innovation in various fields. Medicine and its related fields are typical examples. Because carriers of drugs and genes have to reach the target site, they are required to have a size of nanometer dimension. Also, these carriers have to perform various actions, such as specific binding to target cells, internalization of target cells, drug release, and fusion with cells, just to name a few. Furthermore, they need to guide their contents to reach a target organelle inside a cell. Therefore, it is essential to provide rational functionalities to nanomaterials for effective use in medical fields. Polymer chemistry is one of the best approaches to the production of such nanomaterials with various functionalities. It is expected that polymer-based nanomaterials will be instrumental in the development of new medical technologies.

References

[1] Y. Takakura, M. Hashida, *Crit. Rev. Oncol. Hematol.,* 18, 207-321 (1995).
[2] T. Yamaoka, Y. Tabata, Y. Ikada, *J. Pharm. Pharmacol.* 47, 479-486 (1995).
[3] H. Maeda, Y. Matsumura, *Cancer Res.* 46, 6387-6392 (1986).
[4] H. Maeda, Y. Matsumura, *Crit. Rev. Ther. Drug Carrier System,* 6, 193-210 (1989).
[5] R. Duncan, Nature Reviews, Drug Discovery, 2, 347-360 (2003).
[6] S.E. Stiriba, H. Frey, R. Haag, *Angew. Chem. Int. Edit.* 41, 1329-1334 (2002).
[7] N. Malik, R. Wawattanapatapee, R. Klopsch, K. Lorenz, H. Frey et al. *J. Control. Release.* 65, 133-148 (2000).
[8] M. Ohsaki, T. Okuda, A. Wada, T. Hirayama, T. Niidome, H. Aoyagi, *Bioconjugate Chem.* 13, 510-517 (2002).
[9] M.A. Carnahan, M.W. Grinstaff, *J. Am. Chem. Soc.* 123, 2905-2906 (2001).
[10] K. Kono, C. Kojima, N. Hayashi, E. Nishisaka, K. Kiura, S. Watarai, A. Harada, *Biomaterials,* 29, 1664-1675 (2008).
[11] O.L. Padilla De Jesus, R.H. Ihre, L. Gagne, J.M.J. Fréchet, F.C. Szoka, Jr., *Bioconjugete Chem.* 13, 453-461 (2002)
[12] C. Kojima, K. Kono, K. Maruyama, T. Takagishi, *Bioconjugate Chem.,* 11, 910-917 (2000).

[13] C. Kojima, Y. Toi, A. Harada, K. Kono, *Bioconjugate Chemistry*, 18, 663-670 (2007).
[14] T. Ooya, J. Lee, K. Park, *Bioconjugate Chem.* 15, 1221-1229 (2004).
[15] Y. Haba, C. Kojima, A. Harada, T. Ura, H. Horinaka, K. Kono, *Langmuir,* 23, 5243-5246 (2007).
[16] K. Kono, T. Fukui, T. Takagishi, S. Sakurai, C. Kojima, *Polymer,* 49, 2832-2838 (2008).
[17] C. Kojima, Y. Haba, T. Fukui, K. Kono, T. Takagishi, Macromolecules, 36, 2183-2186(2003).
[18] M. Kimura, M. Kato, T. Muto, K. Hanabusa, H. Shirai, *Macromolecules.* 33, 1117-1119 (2000).
[19] Y. Haba, A. Harada, T. Takagishi, K. Kono, *J. Am. Chem. Soc.,* 126, 12760-12761 (2004).
[20] Y. Haba, C. Kojima, A. Harada, K. Kono, *Macromolecules,* 39, 7451-7453 (2006).
[21] Y. Haba, C. Kojima, A. Harada, K. Kono, *Angew. Chem. Int.* Ed., 46, 234-237 (2007).
[22] K. Kono, T. Miyoshi, Y. Haba, E. Murakami, C. Kojima, A. Harada, *J. Am. Chem. Soc.* 129, 7222-7223(2007).
[23] K. Maruyama, T. Takizawa, T. Yuda, et al., *Biochim. Biophys. Acta.* 1234, 74-80 (1995).
[24] M. Singh, *Current Pharm. Design.* 5, 443-451 (1999).
[25] R.J. Lee, P.S. Low, *J. Biol. Chem.* 269, 3198-3204 (1994).
[26] K. Kono, *Adv. Drug Delivery Rev.* 53, 307-319 (2001).
[27] H. Hayashi, K. Kono, T. Takagishi, *Bioconjugate Chem.,* 10, 412-418 (1999).
[28] K. Yoshino, A. Kadowaki, T. Takagishi, K. Kono, *Bioconjugate Chem.,* 15, 1102-1109 (2004).
[29] K. Kono, R. Nakai, K. Morimoto, T. Takagishi, *FEBS Lett.,* 456, 306-310 (1999).
[30] K. Kono, A. Henmi, T. Takagishi, *Biochim. Biophys. Acta,* 1421, 183-197 (1999).
[31] C.J. Chu, F.C. Szoka, Jr., *J. Liposome Res.* 4, 361-395 (1994).
[32] M. Maeda, A. Kumano, D.A. Tirrell, *J. Am. Chem. Soc.* 110, 7455-7459 (1988).
[33] K. Kono, K. Zenitani, T. Takagishi, *Biochim. Biophys. Acta.* 1193, 1-9 (1994).
[34] K. Kono, T. Igawa, T. Takagishi, *Biochim. Biophys. Acta.* 1325, 143-154 (1997).

[35] K. Kono, Y. Torikoshi, M. Mitsutomi, T. Itoh, N. Emi, H. Yanagie, T. Takagishi, *Gene Therapy*, 8, 5-12 (2001).

[36] N. Sakaguchi, C. Kojima, A. Harada, K. Koiwai, K. Shimizu, N. Emi, K. Kono, *Biomaterials*. 29, 1262-1272 (2008).

[37] N. Sakaguchi, C. Kojima, A Harada, K. Koiwai, K. Kono, *Biomaterials*. 29, 4029-4036 (2008).

[38] E. Yuba, C. Kojima, N. Sakaguchi, A. Harada, K. Koiwai, K. Kono, *J. Controlled Release*, 130, 77-83 (2008).

In: Environmentally Harmonious Chemistry... ISBN: 978-1-60876-428-0
Editors: M. Anpo and K. Mizuno

Chapter 9

An Environmentally-Friendly Plastic Coating Technique Using Gold Nanoparticles for Nanoelectronic Device Applications

Tsutomu Nagaoka and Hiroshi Shiigi

Abstract

In this chapter, new application fields for gold nanoparticles, including electroless gold plating on plastic materials and DNA biosensing technology using a plated array film, are presented. Gold nanoparticles are eco-friendly materials in that they remain stable in harsh chemical environments and have excellent biological compatibility. We applied a nanoparticle-film assembly technique to some electronics applications such as a conductive filler for anisotropic conductive films which is a lead-free conductor system used when massive electrical connections are required between a chip (or a chip-on-board device, etc.) and external input/output lines. The plastic coating technique consists of a single eco-friendly step that does away with the use of toxic chemicals. Also discussed is the application of this film to a DNA biosensor which can detect a single nucleotide polymorphism with a commercially available standard instrument for resistance measurement.

INTRODUCITON

Nanomaterials are considered to constitute a bridge between single molecules and infinite bulk systems, and there are a number of forms that define their structures and properties. These forms include nanoclusters, both artificially and microbally synthesized nanoparticles, quantum dots, core-shell particles, nanowires, and nanotubes [1-3]. In addition to these individual structures, there are also many assembled forms of these materials such as arrays and superlattices. These nanomaterials have been recognized to differ significantly in their physicochemical properties from any single molecules or bulk materials. Along these lines, there has been growing interest in applications of metal nanoparticles to a number of scientific and engineering fields. Among the many different kinds of metal nanoparticles, gold nanoparticles and their assemblies have been used extensively in many areas of chemistry, especially in bio-related chemistry. Gold nanoparticles are very easy-to-handle nanomaterials with respect to stability and the important properties of making the particles useful in such fields are that they have biocompatibility, extremely high extinction coefficients, and straightforwardness in introducing surface functionalities of choice.

Historically, gold colloids were used to obtain glass of various colors such as red and purple. As early as the fourth century AD, colloidal gold was used for decorative purposes, evidenced by the Lycurgus cup made by the Romans [4]. However, the first sound scientific study of gold hydrosols is the paper Michael Faraday published in 1857 [5]. He recognized through his extensive research that hydrosols were composed of divided metal particles. Although much attention has been focused only recently on practical uses, their applications have rapidly expanded to various areas such as catalysis, electronics, surface coating, sensors, and bioanalysis, thanks to the sharp rise of interest in nanotechnology [6-12].

In this chapter, we present a novel and environmentally-friendly coating technique for insulating materials, such as plastic and glass substrates, with gold nanoparticles. The technique has recently been developed in our laboratory by utilizing several unique physicochemical properties of the nanoparticles, aiming for applications in electronics and biosensors [9]. The important advantage of the technique over conventional electroless plastic plating is that our technique does not use toxic chemicals such as chromic acid and other heavy metal ions. In biosensor applications, we also discuss highly sensitive and selective analytical features made up of metal nanoparticle assemblies and superstructures with special emphasis on electronic detection, which does not require any sophisticated instrumentation and chemical systems, such as for optical devices and fluorescent tag molecules [11].

Gold Nano Particle Coating on Plastic Materials

There are growing demands for consumer products using metallized plastics in many engineering fields such as the electronics industry. However, as far as electroplating is concerned, surface coating with a metal on an insulating material has a fundamental problem; electrodeposition of a metal (or electroreduction of metal ion) can only occur on electroconductive materials, which can serve as an electrode. Due to this difficulty, plastic materials have been plated by a different technique called electroless plating. Electroless plating, which has been used for many decades, is a technique to deposit a metal under a mild chemical reducing condition (*see* left side scheme in figure 1). This condition is required to avoid macroscopic metal particle precipitation, which often occurs at high reduction rates. Although this technique is suited for plating on insulating materials such as plastics, special care is required to obtain satisfactory adhesion of a metallic layer to plastics. One such technique is the etching of a surface, in which case surface roughing is necessary to give a micro-structured rough surface to provide a strong enough bonding for subsequent electroless metal deposition since there is no chemical (covalent) bonding expected between the plastic surface and metallic layer. This process, developed in the mid-1960's, relies on chromic-acid based oxidation and the current plating technologies on plastics still employ this toxic method developed at that time. After surface roughing, the plastic surface is activated with a solution containing tin and palladium salts. This activation process gives a deposition of palladium nuclei that catalyzes subsequent electroless nickel or copper growth. Once the plastic material has been coated with an electrically conductive copper or nickel surface layer, it can subsequently be electroplated using conventional electroplating technology.

As seen in figure 1, the whole process consists of many chemical steps. Moreover, the technology relies heavily on toxic compounds such as chromic acid and heavy metals which cause waste disposal problems. As a result, each procedure in this technology can increase environmental loads and give rise to serious pollution. Along with these problems, the recipes involved in plating solutions are often comprised of jealously guarded commercial secrets and these formulations may contain unexpected ingredients [13] which may lead to further environmental problems.

Recently, we have developed a straightforward plastic coating technique based on metal nanoparticle deposition [9, 14, 15]. The gold nanoparticles used in these experiments are prepared in an aqueous media with ascorbic or citric acid as the reducing agent. The gold particles and plastic substrate are then agitated in an aqueous linker (alkanethiol) suspension. Figure 1 compares the experimental procedures of our technique with a standard electroless plastic plating technique. As can be readily observed, our technique is much more environmentally friendly, doing away with the use of toxic chemicals such as chromic acid, and is comprised of far fewer steps than the electroless technique. In our technique, the possible wastes are Cl^-, citric or ascorbic acid which have been used as food additives and small amounts of alkanethiol which can be easily separated from water. The plating solution becomes completely clear in color after finishing the plating, meaning that the transfer of gold to the plastic is quantitative.

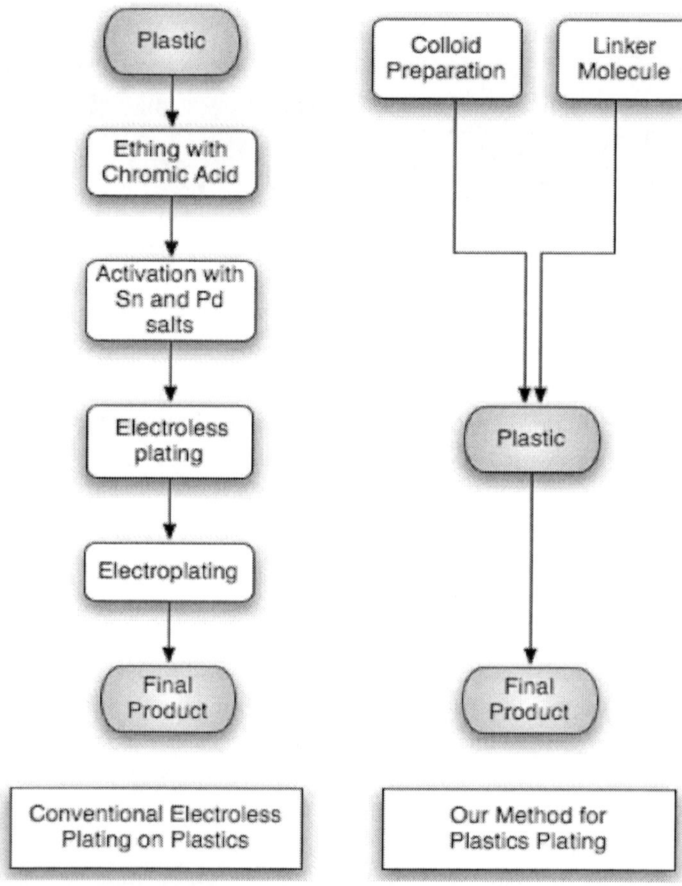

Figure 1. Comparative Schemes for the plastic coating techniques.

The whole process of this nanoparticle plating is illustrated in figure 2 which is explained as follows: First, chloroauric acid is reduced with citric acid to produce a gold nanoparticle with a diameter of 46±25 nm.

$$n\mathrm{AuCl_4^-} + 3n\,\mathrm{e^-} \rightarrow \mathrm{Au_n}\ \text{(gold colloid)} \qquad (1)$$

Then, the particle is combined with a polystyrene plate in the presence of alkanethiol in an aqueous solution.

$$\mathrm{Au_n} + m\mathrm{C_4H_9SH} \rightarrow \mathrm{Au_n(SC_4H_9)_m} + m/2\ \mathrm{H_2} \qquad (2a)$$

$$\text{Polystyrene} + \mathrm{Au_n(SC_4H_9)_m} \rightarrow \mathrm{Au_n(SC_4H_9)_m} : \text{Polystyrene} \qquad (2b)$$

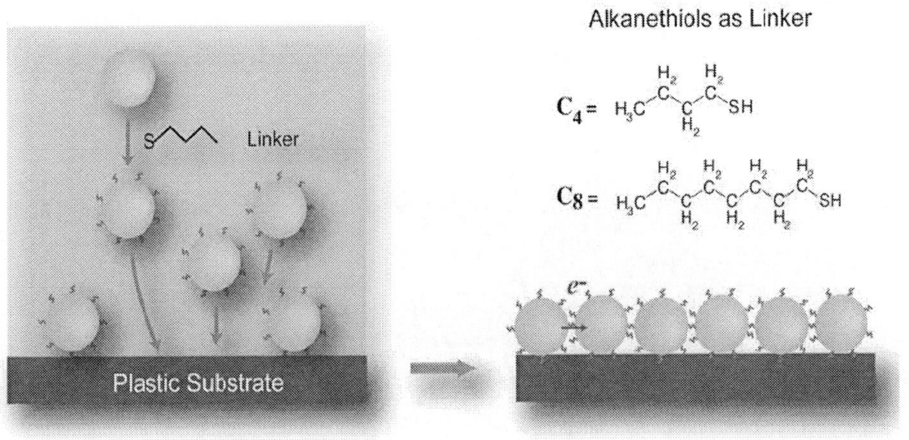

Figure 2. A model illustration of the nanoparticle-based conductive plastic coating.

The reactions 2a and 2b occur in the solution at the same time. Reaction 2a is actually an exchange reaction with respect to a protecting molecule: Initially, a gold nanoparticle prepared with organic acid adsorbs its dissociated form (acid anion), which eventually gives a negative charge on the particle surface, avoiding aggregation of each particle in solution by an electrostatic repulsive force. Though not shown explicitly, the acid anion is replaced with thiol in Reaction 2a (in our typical recipe, the Au/thiol molar ration is 7.7/1). The hydrophobic nature of the thiol-protected gold nanoparticles leads to their bonding to the surface of plastic, such as polystyrene, to create a lustrous and robust gold film on the plastic. The film-plastic bonding is strong enough to pass the peel-off tape test. However, it is evident from the SEM image (figure 3) that the plastic is densely tiled with nanoparticles. This film consists of 3 stacked particle layers on average (~150 nm in thickness). Surface roughing of the plastic was observed by AFM when it was treated only with butanethiol and this probably occurs by the interaction of the hydrophobic end (CH_3-) of the thiol with the plastic surface in the plating stage [9].

Figure 3. A SEM image of the conductive nanoparticle film prepared on a polystyrene substrate (linker, butanethiol; particle size, 46±25 nm).

Interestingly, a strong dependence was found when the resistivity of the coated film was plotted against the molecular length of the alkanethiols. The resistance of the film prepared by butanethiol (2.1×10^{-6} Ω cm) was close to that for metallic gold (1.3×10^{-6} Ω cm), but the film prepared with the longer thiols makes its resistance higher (figure 4). The film conductance for butanethiol is one of the highest values among the reported nanoparticle films in literature [14-17]. With this film, the tunneling barrier between two particles determines the local conductivity so that the linker length has a crucial effect in the conductivity. The conductivity (σ) between the isolated metal particles due to tunneling electron transfer is proportional to $\exp(-\beta x)$, where β is the electronic coupling factor and x the distance separating the neighboring two particles [18-21]. A β-value of 0.6, which was evaluated from the σ—x plots by using an x value of 2×(thiol length), was smaller than the typical value for an alkanethiol-crossing electron transfer (~1 $Å^{-1}$) [18, 21]. The smaller value implies that a high degree of interdigitation of the hydrocarbon chains from the neighboring particles may occur. Such interdigitation, probably due to strong hydrophobic interactions between the hydrocarbon chains, has also been reported for alkanethiol separated nanoparticle arrays formed on glass substrates [21]. The strong coupling between the particles is, thus, an important factor in maintaining both the robustness and conductivity of the film as well as the bonding of the particle with the plastic surface [14].

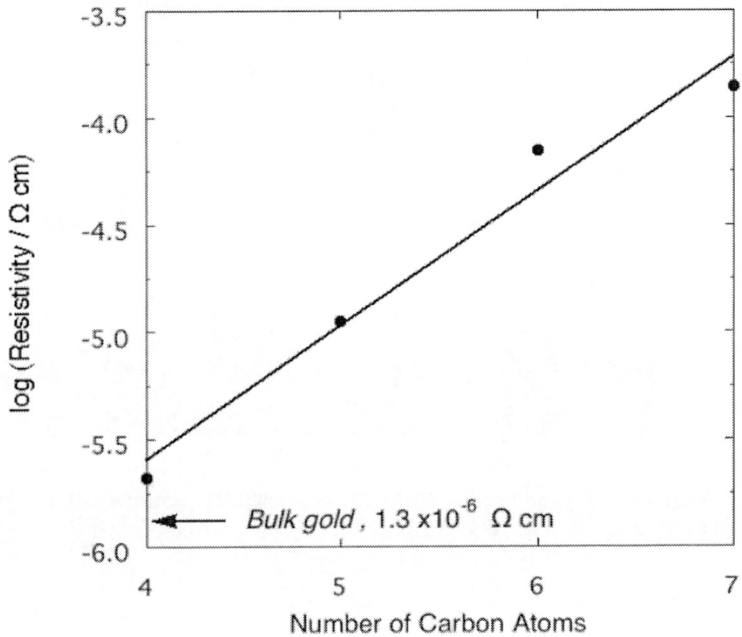

Figure 4. Resistivity of the gold nanoparticle films prepared on a polystyrene plate as a function of the linker length (number of carbon atoms in the alkanethiols).

One important advantage of wet chemical plating techniques as compared to evaporation based techniques is that the former can coat various types of curved surfaces such as the inside wall of tubing and particle surfaces. We show here some of the results for the latter case (figure 5). As an application of such particles, we have investigated their electric properties, aiming at a conductive filler for an anisotropic conductive film (ACF). ACF is a lead-free and environmentally-friendly epoxy-based conductor system which has been used for many years to make local electrical connections between a liquid-crystal-display panel unit and external input/output (I/O) lines or for other places where massive electrical connections are required between a chip (VLSI, a chip-on-board device, etc.) and external I/O lines. ACF usually consists of conducting particles dispersed in an insulator matrix (epoxy resin, etc.), and the former creates a local conducting path only between two metallic terminals (e.g. between a die bump of a flip-chip device and an I/O terminal) when pressed on implementation of the devices. The resistance of this particle was 0.1–0.3 Ω / particle.

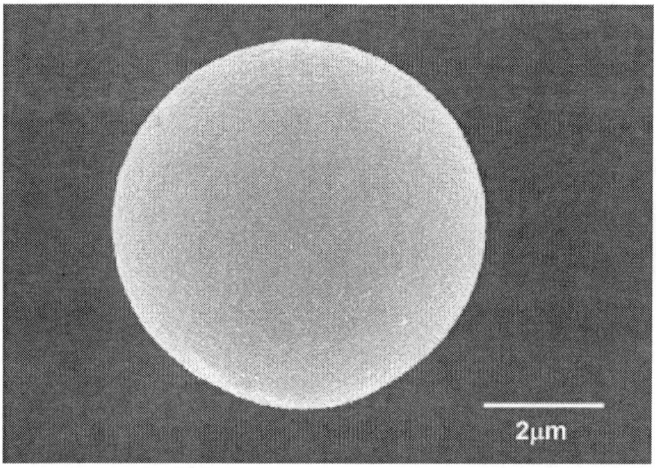

Figure 5. A 5-μm polyacrylate bead coated with a butanethiol-protected 30-nm Au particle.

Another potential application for the nanoparticle-based system is the construction of superstructures that have recently been reported by a number of authors [3]. In this connection, there is keen interest in building 1–3 D nanostructures by using functionalized nanoparticles as a future technology for manufacturing electronically/optically coupled nanodevices. One such example is the 1D-nanoparticle based system required for high-throughput molecular-level computing. For this purpose, DNA computing has been proposed as bio-stimulated computing devices and the simplest case for such systems is the DNA-templated nanowire or 1D nanoassembly [22, 23]. Naturally occurring DNA is one of the potential source materials for the fabrication of such structures from the viewpoint of its biological abundance. Figure 6 shows an example of a DNA-templated 1D-structure assembled on a plastic substrate using positively charged 2-nm Au nanoparticles which were synthesized in the authors' laboratory [24, 25]. Since DNA is a poly-phosphate anion (*see* figure 6 B), a positively charged nanoparticle is required for the creation of such a structure through an electrostatic interaction. Although most of the gold nanoparticles reported so far have negative charges, we have successfully prepared such particles in a very straightforward fashion [25, 26]. The precise physicochemical properties of this structure, such as electronic conductance and optical properties, are presently under investigation.

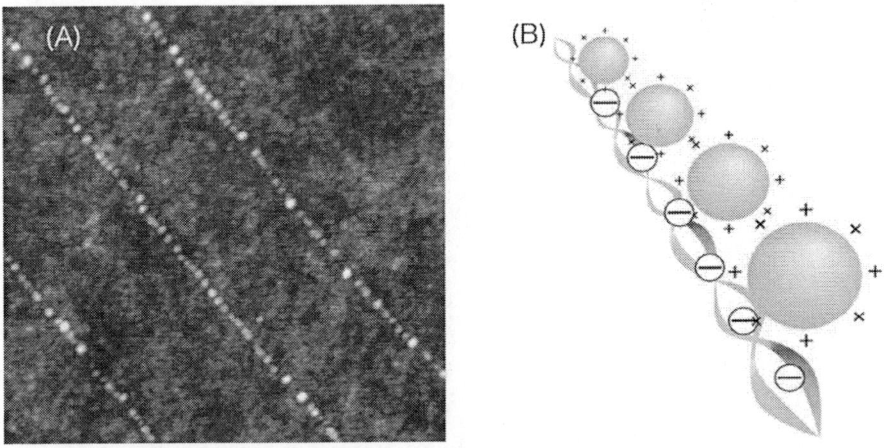

Figure 6. (A) An AFM image of a DNA-templated 1D-nanostructure using positively charged 2-nm gold particles (courtesy of Dr. H. Nakao), and (B) a schematic drawing.

NANOPARTICLE ARRAY BASED BIOSENSORS

As discussed above, the resistance of a nanoparticle film can easily be controlled by selecting the linker molecule. Low resistance films prepared with shorter linker molecules can find various applications in the electronics industry, however, films possessing higher resistance also have usefulness in many application fields. One such example is their application to chemical sensors of various types.

Gas sensors called "Chemiresisters", which are fabricated from electrically conducting particles (carbon black, etc.) dispersed in an insulating matrix (polyethylene, etc.), have been used for many years [27,28]. The insulator matrix is used to give a moderately high film resistance (100 kΩ – 1 MΩ) for sensitive measurements with commercially available standard equipment. In a similar idea, the use of particle array films has recently been reported to detect hydrophobic gases by monitoring resistance changes through its adsorption to the array film (figure 7). In this sensor arrangement, the insulating matrix is not used but the protecting layer on the nanoparticle serves this purpose instead [10, 20, 21, 29, 30]. Naturally, such a layer (alkane- or aromatic-thiols) favors the adsorption of hydrophobic gases (toluene, tetrachloroethylene) more than that for hydrophilic ones (water, and propanol), and the film resistance changes almost linearly with the vapor pressure. The working mechanism is not very clear with these types of sensors, but changes in the dielectric property between the particles due to gas adsorption are considered to be responsible [19, 20]. It was also reported that copper ions introduced in the array reduced the hydrophobicity of the array and ethanol was determined in this arrangement [21].

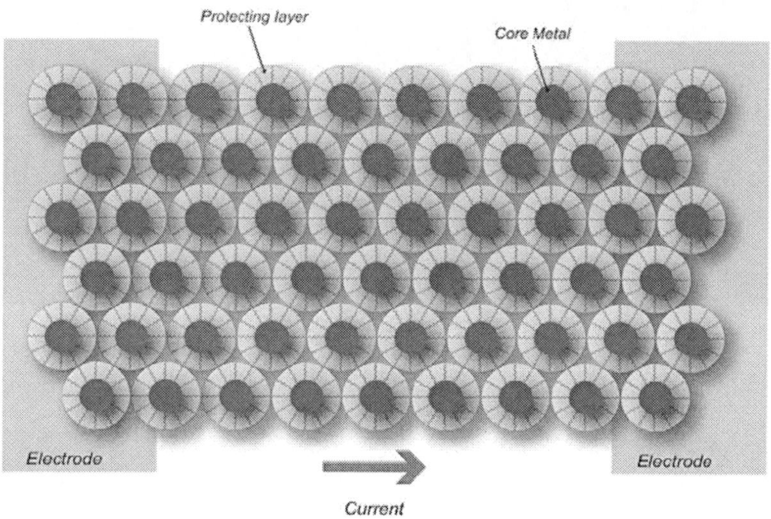

Figure 7. A model illustration of 2D nanoparticle array film prepared on an insulating material: each particle is separated from its neighboring particles with a protective thiol layer, which acts as a tunneling barrier for electron migration.

The nanoparticle array is a potential candidate for nanosized sensors based on a molecular device design (figure 8 A). A molecular electronic device is the ultimate goal of such small-sized electronic components such as nano-FETs and is created between two metallic terminals arranged with a molecular size gap (~1 nm). A molecule of choice, exhibiting active electronic properties, is inserted between the terminals. If a receptor molecule, which can recognize its guest molecule through the complex formation, is placed appropriately, it can constitute the smallest chemical sensor possible. Such nanosized sensors are currently receiving keen attention towards high-throughput chemical data acquisition when assembled in a large-scale array system. Although creating a 1-nm gap still involves technical difficulties such as reproducibility, several laboratory level techniques are now available (figure 8) [31]. In this connection, we have applied the nanoparticle array assembly technique for such a purpose (figure 8. D). This technique, as already mentioned above, can easily set up a small gap by self-assembly and we have applied this technique to DNA sensing [11, 32, 33].

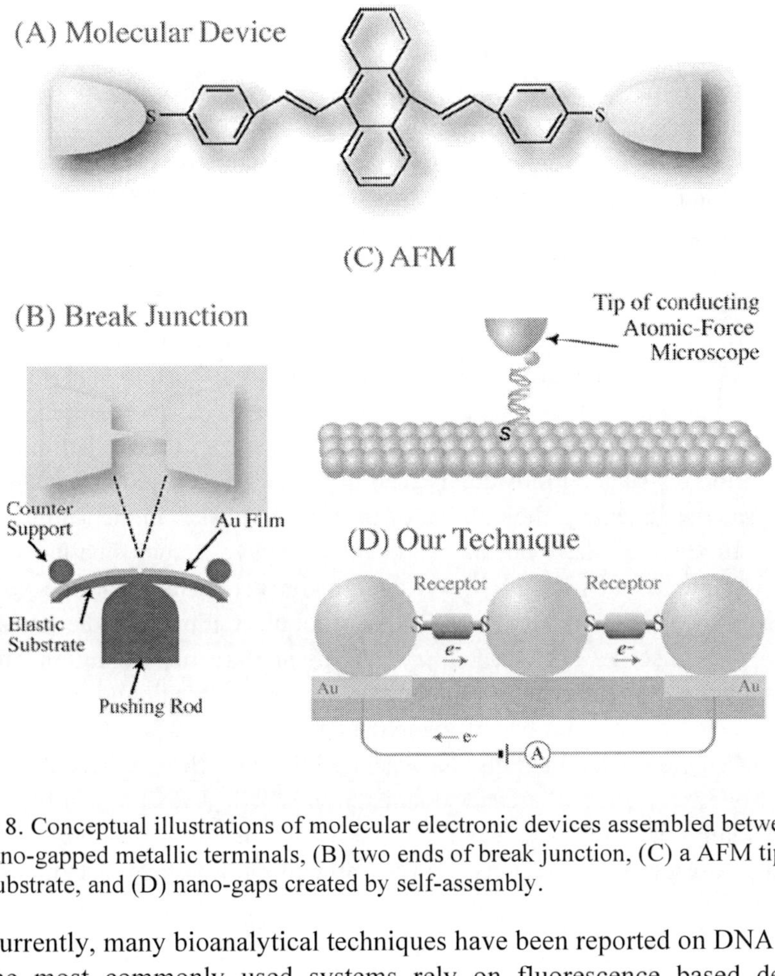

Figure 8. Conceptual illustrations of molecular electronic devices assembled between (A) two nano-gapped metallic terminals, (B) two ends of break junction, (C) a AFM tip and gold substrate, and (D) nano-gaps created by self-assembly.

Currently, many bioanalytical techniques have been reported on DNA sensing but the most commonly used systems rely on fluorescence based detection. However, such systems use sophisticated optical systems which make miniaturization of the instrument difficult. Furthermore, the system works with elaborately designed fluorescent tag molecules, making handling of the sample difficult and analysis expensive. However, a system based on conductivity measurements can solve such problems because of the straightforwardness in the system design and measurements. Figure 9 shows the experimental procedures for assembling a nano-gapped array film on a glass substrate. Interdigited Pt electrodes (a), each consisting of 60 fingers with a finger-to-finger distance of 5 μm, were first immersed into a decanedithiol linker solution to modify the Pt sections with the linker. Afterwards, the electrode assembly was submerged into an 80-nm gold nanoparticle dispersion to anchor the particle on the Pt electrodes. This decanedithiol-nanoparticle submergence cycle was repeated until the nanoparticles covered the entire glass surface (figure 9 g and h). Thus produced ready-for-use films gave resistance values ranging from 200-1000 ohm, depending on the number of submergence cycles applied.

There has been much controversy about the conductivity of DNA [34-36], and recent studies using AFM and STM techniques reported that short oligonucleotides have measurable conductivities (10–100 M) [37, 38].

Based on this array film, we have incorporated a DNA sensing ability by immobilizing a 12-mer thiol-modified oligonucleotide on the nanoparticle as a probe (figure 10 B). To measure a DNA hybridization event, a 5-μL pH-7.4 buffer was first applied on the array to monitor the control current and the signal response at this stage was very stable. Then, a 5-μL 12-mer oligonucleotide sample dissolved in the buffer was spiked onto the sensor surface. An immediate decrease in the resistance followed, as seen in figure 10 A. The magnitude of the response was dependent on the number of mismatched bases in the sample strand and, interestingly, the 1-base mismatched strand (c) gave a much smaller response than the complementary strand (d). Although the mechanism for the decrease (higher resistance for the mismatched DNA) is not clear at present, the unmatched base pair may disorder the stacked π-overlapping structure along the double strands and work as a defect in the charge transfer path. The nonlinear dependence of the resistance on the number of mismatched base pairs can be applied to the detection of single nucleotide polymorphisms (SNP), which is represented by the change of one nucleotide in a DNA double strand. SNP detection can be used to diagnose many diseases and to study variations in drug responses among individuals. Currently, DNA microarrays based on fluorescent detection are a most widely used technological platform that enables SNP detection in the area of pharmacogenetic diagnosis. Therefore, the simple detection mechanism of our method can provide a more accurate and cost-effective way to perform such detection. At present, the detection limit of this technique is ~50 fmol for 24-mer oligonucleotides [33].

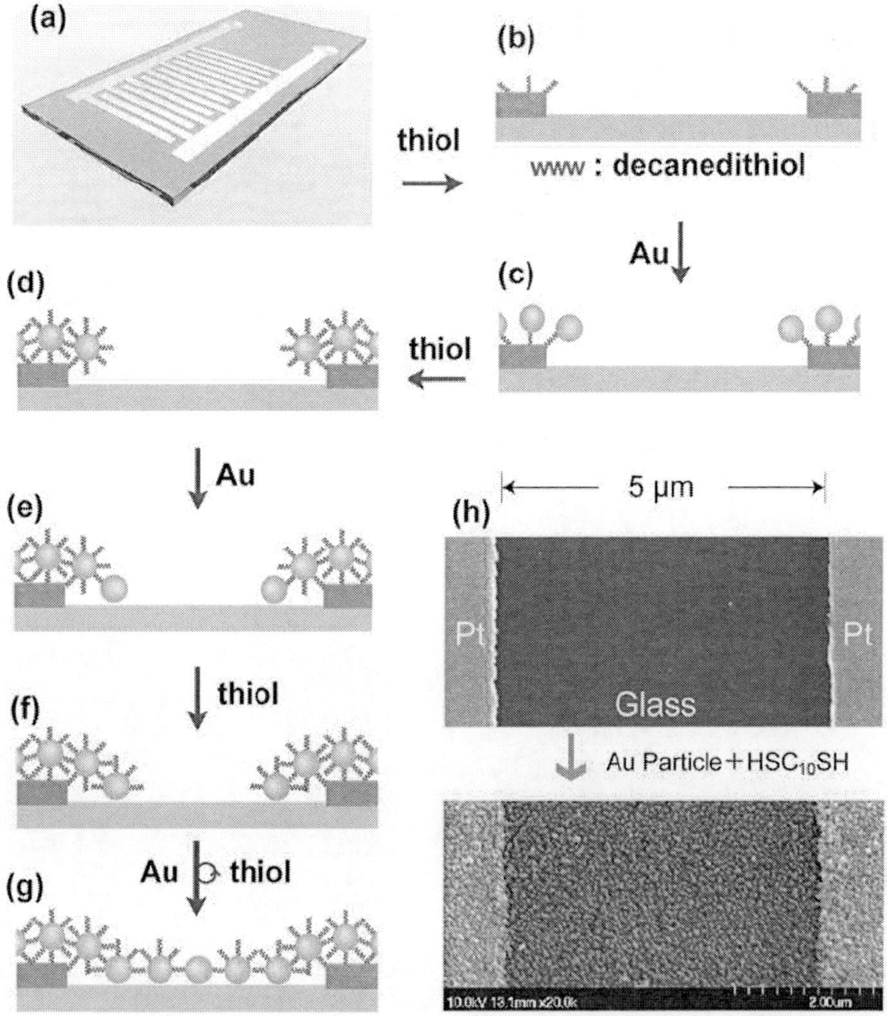

Figure 9. Assembling of nano-gapped nanoparticle array film on interdigited electrodes: (a) a model illustration of the interdigited electrodes (one pair of Pt electrodes on the chip; each row consists of 65 fingers; separation of the finger, 5 μm; finger width 10 μm); (b) – (f), gold nanoparticle deposition procedures by self-assembly; (g) the final product; (h) SEM images of the electrode before and after deposition.

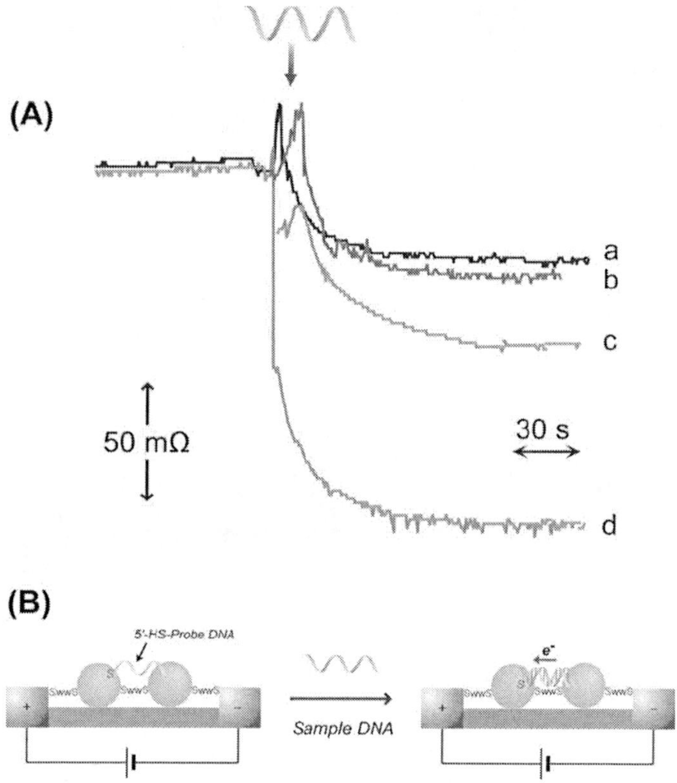

Figure 10. (A) Resistance decreases on the addition of 4 different target 12-mer oligonucleotide on the 80-nm gold nanoparticle array film, and (B) the detection protocols for hybridization event of a sample with 12-mer oligonucleotide probe immobilized on the gold nanoparticle. The sample DNA contains (a) 11-base mismatches, (b) 4-base mismatches, (c) 1-base mismatch, (d) 0-base mismatch (complementary strand).

CONCLUSION

In this chapter, we have presented some novel approaches in electroless conducting plating and biosensing technologies by using gold nanoparticle array films which are derived from similar compositions but possess different electrical characteristics. Thus, by selecting only the linker molecule, this film assembling technique provides a wide variety of applications ranging from electronics to biosensing.

The gold nanoparticle is an eco-friendly material in that it remains stable even under harsh chemical environments and has excellent biological compatibility. An attempt has been made to use this material in electroless plating and its coating protocols were compared with a conventional electroless process, showing that the nanoparticle 2D coating technique requires much fewer steps to obtain metal coating on plastic materials. The electrical property of this film can easily be controlled with the length of the linker molecule and is comparable to bulk gold conductivity when a short thiol is used. Films prepared with longer linker molecules show higher resistance, allowing us to apply such films to a conductivity-based chemical sensor by installing a receptor between the particles. We have also demonstrated that this simple sensor has very high sensitivity and selectivity with an oligonucleotide probe, satisfying the practical requirement in SNP diagnoses.

REFERENCES

[1] Rao, C. N. R.; Müller, A.; Cheethanm, A. K. (Ed.) The Chemistry of Nanomaterials; *WILEY-VCH: Weinheim,* 2004; Vol. 1 & 2.

[2] Sastry, M.; Ahmad, A.; Khan, M. I. Kumar, R. In Nanobiotechnology; Niemeyer, C. M.; Mirkin, C. A.; Ed.; *WILEY-VCH: Weinheim,* 2004; 126-145.

[3] Kotov, N. A. Nanoparticle Assemblies and Superstructures; CRC Press: Boca Raton, FL, 2006.

[4] Brust, M.; Kiely, C. J. In Colloids and Colloid Assemblies; Caruso, F.; Ed.; *WILEY-VCH: Weinheim,* 2004; 96-119.

[5] M. Faraday: *Philos. Trans. R. Soc.* London: 147, 145(1857).

[6] Haruta, M.; Yamada, N.; Kobayashi, T. J. *Catal.* 1989, 115, 301-309.

[7] Kotov, N. A.; Tang, Z. In Nanoparticle Assemblies and Superstructures; Kotov, N. A.; CRC Press, Boca Raton, FL, 2006; 3-73.

[8] Feldheim, D. L.; Keating, C. D. *Chem. Soc. Rev.* 1998, 27, 1-12.

[9] Shiigi, H.; Yamamoto, Y.; Yakabe, H.; Tokonami, S.; Nagaoka, T. *Chem. Commun.* 2003, 1038-1039.

[10] Wohltjen, H.; Snow, A. W. *Anal. Chem.* 1998, 70,2856-2859.

[11] Shiigi, H.; Tokonami, S.; Yakabe, H.; Nagaoka, T. *J. Am. Chem. Soc.* 2005, 127, 3282-3281.

[12] Mirkin, C. A.; Letsinger, R. L.; Mucic, R. C.; Storhoff, J. J. *Nature.* 1996, 382, 607-609.

[13] Oldham, K. B.; Myland, J. C. *Fundamentals of Electrochemical Science;* Academic Press: San Diego, CA, 1994; 96-101.
[14] Yamamoto, Y.; Shiigi, H.; Nagaoka, T. *Electroanalysis*. 2005, 17, 2224-2230.
[15] Yamamoto, Y.; Shiigi, H.; Nagaoka, T. *Solid State Ionics*. 2006, 177, 2325-2328.
[16] Musik, M. D.; Keating, C. D.; Keefe, M. H.; Natan, M. J. *Chem. Mater.* 1997, 9, 1449-1501.
[17] Liu, Y.; Wang, Y.; Claus, R. O. *Chem. Phys. Lett.* 1998, 298, 315-319.
[18] Bard, A. J.; Faulkner, L. R. *Electrochmical Methods;* WILEY: Hoboken, NJ, 2001, 624-627.
[19] Neugebauer, C. A.; Webb, M. B. *J. Appl. Phys.* 1962, 33, 74-82.
[20] Evans, S. D.; Johnson, S. R.; Cheng, Y. L.; Shen, T. J. *Mater. Chem.* 2000, 10, 183-188.
[21] Zamborini, F. P.; Leopold, M. C; Hicks, J. F.; Kulesza, P. J.; Malik, M. A.; Murray, R. W. *J. Am. Chem. Soc.* 2002, 124, 8958-8964.
[22] Keren, K.; Sivan, U.; Braun, E. In Bioelectronics Willner, I.; Katz, E.; Ed.; WILEY-VCH: Weinheim, 2005; 265-285.
[23] Braun, E.; Eichen, Y.; Sivan, U.; Gen-Yoseph, G. *Nature.* 1998, 391, 775-778.
[24] Nakao, H.; Gad, M.; Sugiyama, S.; Otobe, K.; Ohtani, T. *J. Am. Chem. Soc.,* 2003, 125, 7162-7163.
[25] Nakao, H.; Shiigi, H.; Yamamoto, Y.; Tokonami, S.; Nagaoka, T.; Sugiyama, S.; Ohtani, T. *Nano Lett.,* 2003, 3, 1391-1394.
[26] Shiigi, H.; Yamamoto, Y.; Yoshi, N.; Nakao, H.; Nagaoka, T. *Chem. Commun.* 2006, 4288-4289.
[27] Lundberg, B.; Sundqvist, B. *J. Appl. Phys.* 1986, 60, 1074-1079.
[28] Lonergan, M. C.; Severin, E. J.; Doleman, B. J.; Beaber, S. A.;Grubbs, R. H.; Lewis, N. S. *Chem. Mater.* 1996, 8, 2298-2312.
[29] Han, L.; Daniel, D. R.;Maye, M. M.; Zhong, C.-J. *Anal. Chem.,* 2001, 73, 4441-4449.
[30] Krasteva, N.; Besnard, I.; Guse, B.; Bauer, R. E.; Müllen, K.;Yasuda, A.; Vossmeyer, T. *Nano Lett.* 2002, 2, 551-555.
[31] Mayor, M.; Weber, H. B.; Waser, R. In Nanoelectronics and information technology; Waser, R.; Ed.; *WILEY-VCH: Weinheim,* 2003,503-525.
[32] Tokonami, S.; Iwamoto, I.; Hashiba, K.; Shiigi, H.; Nagaoka, T. *Solid State Ionics,* 2006, 177, 2317-2320.
[33] Tokonami, S.; Nishide, Y.; Shiigi, H.; Nagaoka, T. *BUNSEKI KAGAKU* in press.

[34] Murphy, C. J.; Arkin, M. R.; Jenkins, Y.; Ghatlia, N. D.; Bossmann, S. H.; Turro, N. J.; Barton J. K. *Science.* 1993, 262, 1025-1029.

[35] Porath, D.; Bezryadin, A.; de Vries, S; Dekker, C. *Nature.* 2000, 403, 635-637.

[36] Wangenknecht, H.-A. Charge Transfer in DNA; *WILEY-VCH: Weinheim,* 2005.

[37] Sumaoka, J.; Pan, F. Nonaka, A.; Takeuchi, O.; Shigekawa, H.; Komiyama, M. *Chem. Lett.* 2004, 33, 700-701.

[38] Wierzbinski, E.; Arndt, J.; Hammond, W.; Slowinski, K. *Langmuir.* 2006, 22, 2426-2429.

In: Environmentally Harmonious Chemistry... ISBN: 978-1-60876-428-0
Editors: M. Anpo and K. Mizuno

Chapter 10

POLYSILANES: NOVEL SIGMA-CONJUGATED POLYMERS

Kunio Oka

ABSTRACT

Polysilanes are polymers of Si-catenated backbone and each silicon atom bares two organic side groups. The sigma-electrons of the Si-Si bonds delocalize over the long range of the backbone (sigma-conjugation) resulting in unusual physical properties such as strong absorption at 300-400 nm and efficient fluorescence in the ultraviolet region, high conductivity upon doping with oxidants, high photoconductivity and high NOL. Polysilanes have, therefore, been studied as a photoconductor, emitting material and/or carrier transporting materials of LEDs, photoresists, and photo-induced radical initiators. Organic side groups control the conformation of the main chain which controls the conjugation efficiency. The conformation of polysilanes is sensitive to the circumstances of the polymer and, thus, brings about various chromotropisms such as thermo-, piezo, electro-, iono- and solvato-chromisms.

1. INTRODUCTION

Polysilanes are polymers having a catenated silicon backbone and two organic groups are attached to each silicon atom. (figure 1). Although they should be called polysilylenes, "polysilanes" is commonly used . Homopolymer and copolymers are represented as (1) and (2) where both organic groups are either the same or different. Polysilanes composed from disilylene repeating units (3) were also prepared. Unlike structurally resembling carbon polymers, polysilanes exhibit unusual electronic and optical properties such as strong absorption and efficient emission in the UV region, high photoconductivity, and high non-linear optical properties. These novel properties make polysilanes an interesting new material for various technological applications. This chapter deals with an outline of polysilane chemistry and some of its technological applications.

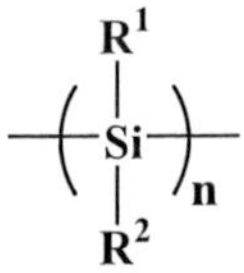

(1)

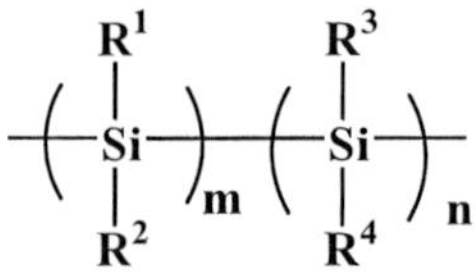

(2)

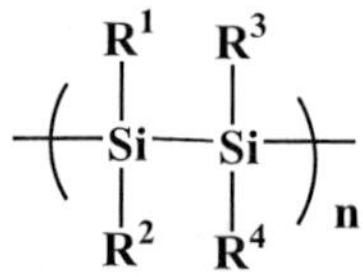

(3)

2. SYNTHESIS

Carbon polymers are prepared using a wide variety of starting materials (monomers) and various polymerization methods (radical, ionic, group, or atom-transfer, polysilanes, ROP, condensation, living and stepwise, etc.). However, polysilanes can be obtained utilizing only limited kinds of monomers; halosilanes(X_nSiR_{4-n}), organosilane(R_nSiH_{4-n}), and some Si-Si bonded compounds. This is due to the limited availability of unsaturated silicon compounds, disilenes($R_2Si{=}SiR_2$), in practical quantities and with enough stability to be stored and handled in laboratory work. Thus, polysilanes have been prepared mainly by the following three methods.

1. Reductive coupling reaction of dichlorodiorganisilanes.
2. Dehydrogenative coupling reaction of organosilanes
3. Polymerization of reactive Si-Si compounds including “Masked disilenes" and strained-ring compounds.

2.1. Reductive Coupling Reaction of Dichlorodiorganisilanes

The reductive coupling reaction of dialkyldihalosilane, R_2SiX_2, with alkali metals (Wurtz-type reaction) (eq. 1) is, up to now, the major and general method to obtain a wide variety of polysilanes in practical quantity. Dialkyldichlorosilanes are usually added into a refluxed toluene containing a fine sodium dispersion[1]. The yield and molecular weight distribution of the products are highly dependent on the reaction condition and the nature of the organic groups on the silicon atoms. For the same monomer, the reaction temperature, concentration of the monomer, reaction sequence, solvent, catalyst and/or additives are known to affect the reaction. The reaction of sodium with dichlorodialkylsilanes in refluxed toluene generally gives satisfactory results.

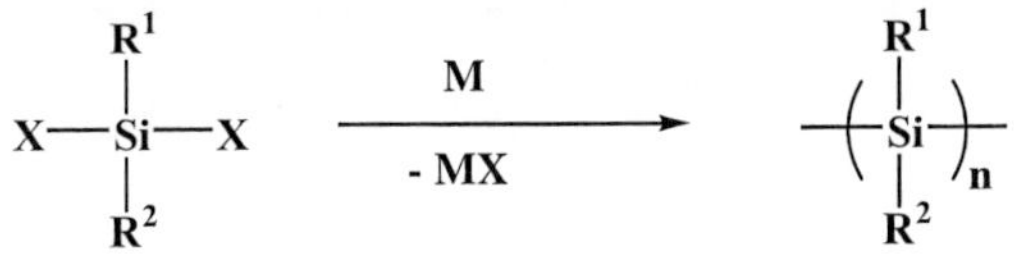

(1) (1)

Dibromo-derivatives are generally more reactive than dichloro-derivatives. Potassium, Na/K alloy, or potassium-graphite are used for the polymerization of less reactive monomers. Additives such as ethyl acetate[2] or the Cu-catalyst[3] improves both the yield and molecular weight distribution. Inverse addition [4] (i.e., the addition of a sodium dispersion into a monomer solution) and a low temperature reaction [5] are also effective to improve both the molecular weight and yields. Ultrasound agitation[5,6] and the addition of a phase transfer catalyst (crown-ether or cryptande)[7] improves the yields with a compensation of the molecular weight. Electrochemical condensation using magnesium electrodes has also been reported[8].

The reductive coupling reaction usually gives polysilanes with polymodal molecular weight distribution. High ($M>10^5$), medium ($M= -10^4$) and low ($M= -10^3$) fractions are generally detected in GPC. A low molecular weight fraction consists of linear and cyclic oligomers. Oigomers usually involve Si-H end groups. High molecular polysilanes are generally easily isolated by fractional precipitation by adding 2-propanol into a toluene solution of polysilanes. The reaction is believed to proceed via an anion-radical intermediate, as depicted in Scheme 1[9].

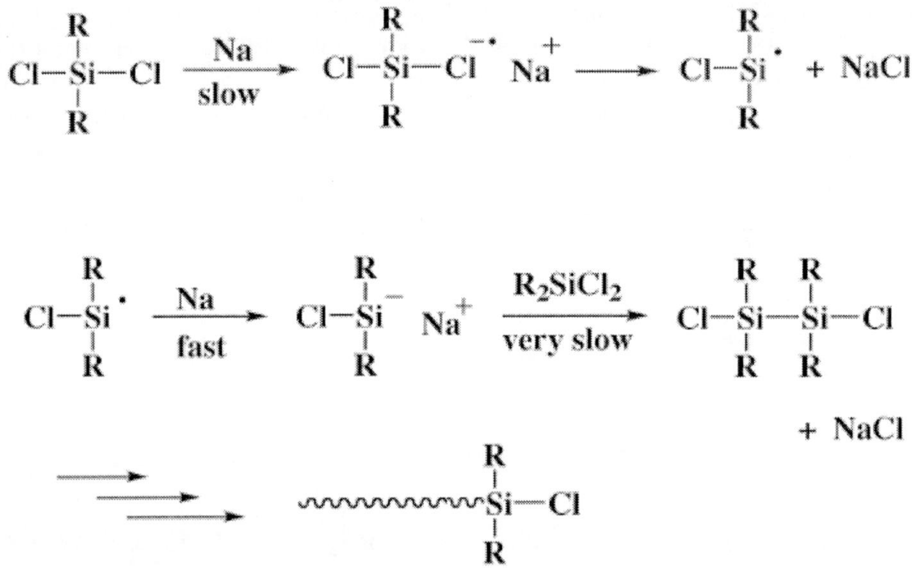

Scheme 1.

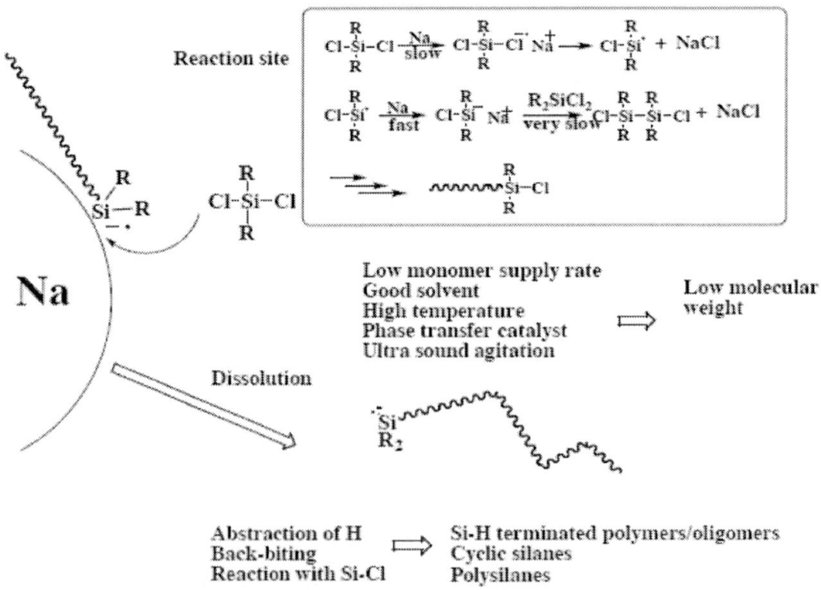

Figure 1. Polymerization reaction of dichlorodiorganosilanes on a sodium surface.

The condensation reaction proceeds mainly on the surface of alkali metals, as depicted in figure 1. When dichlorosilane monomers are delivered to the site where a silyl anion radical is formed, a Si-Si bond is formed and generates disilyl anion radicals. Repetition of this sequence leads to anion radicals of a longer chain on the metal surface. A low monomer supply rate, high temperature, highly diluted reaction conditions or the use an effective solvent causes the growing polysilane moieties to diffuse into the solvents. The diffused reactive intermediates will attack the Si-Cl groups or abstract hydrogen atoms from the solvent, resulting in polysilanes of low molecular weight. Thus, a low molecular weight fraction, especially an oligomer, contains Si-H groups. The intermediates also attack Si-Si bonds to bring about cyclic oligomers or the scission of the main chain. Cyclic oligomers are also the main component of the low molecular fraction. The diffusion of the reactive intermediates leads to a fresh metal surface (reaction sites) and a high yield of relatively low molecular polysilane can be expected. A higher rate of the monomer supply, lower reaction temperature, or poor solvent system will prevent diffusion of the reactive intermediates into the solvent and this favors the propagating process and enhances the molecular weight of the polysilanes. Highly grown intermediates will wrap-up the metal surface and prevent further delivery of the monomer.

Copolymers (2) and (3) are prepared similarly by the co-condensation reaction of two monomers and the condensation of 1,2-dihalosilanes, respectively.

2.2. Dehydrogenative Coupling Reaction of Organosilanes

Dehydrogenative coupling of organosilanes with transition metal complexes affords safer and more easily controlled reactions than the reductive coupling reaction of dihalosilanes [eq. 2].

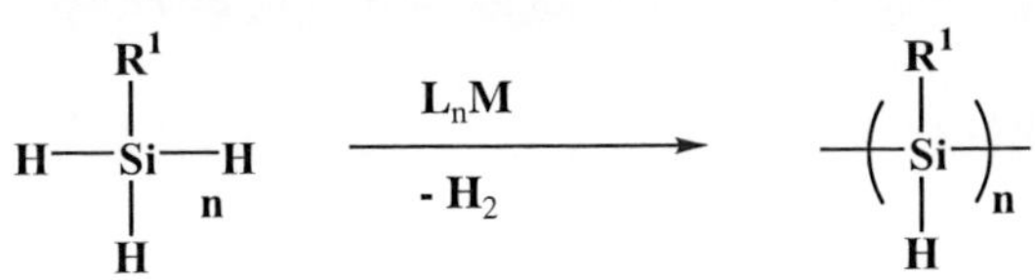

(4) (2)

Metalocenes of Group IV elements[10] are the most effective in yielding polysilanes among the diverse metal complex examined. Unfortunately, polysilanes with satisfactory molecular weight are obtained only by using arylsilane monomers ($ArSiH_3$). Alkylsilanes ($RSiH_3$) affords oligomeric products while dialkylsilane(R_2SiH_2) undergoes only dimerization. Dehydrogenative polymerization is believed to proceed *via* a sigma-bond metathesis mechanism, as shown in Scheme 2[11]. Moreover, the nucleophilic reaction of the Si-H bond-containing products affords new polysilanes with various functional groups[12].

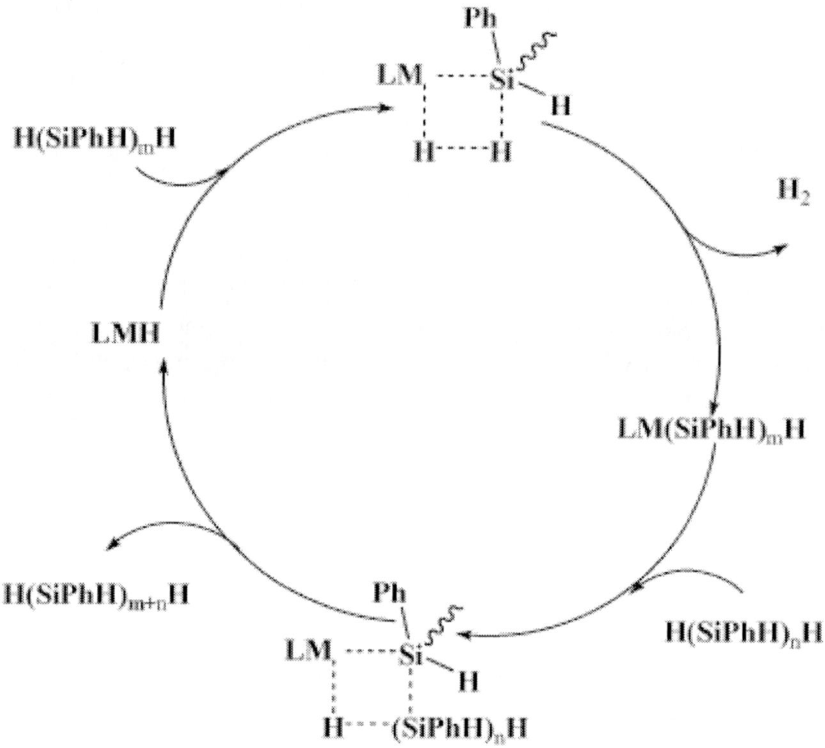

Scheme 2.

2.3. Other Polymerization Methods

Disilenes ($R_2Si=SiR_2$) are not used as the monomer for a polymerization reaction because of limitations in affordability. Instead, bicyclo-compounds obtained by the reaction of 1,2-dichlorodisilanes with biphenyl anion radicals behave as the equivalent of disilenes and are referred to as "Masked disilenes"[13]. Bicylo-compounds undergo living polymerization by the action of the anionic initiator (eq. 3). This smart method affords regioselective polysilanes with small polydispersity. Applicability of this polymerization procedure depends on the availability of the monomers. Aryl-substituted masked disilene suffers from this living polymerization reaction.

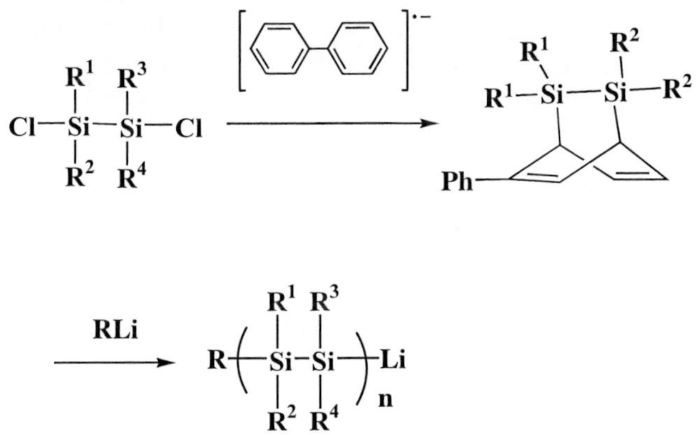

(3)

Because of the poor availability of the required strained ring compounds, only a limited number of cyclic silanes were studied for ring opening polymerization (eq. 4). Some cyclotetrasilanes[14,15], and cyclopentasilanes[16] were found to undergo ROP to give polysilanes.

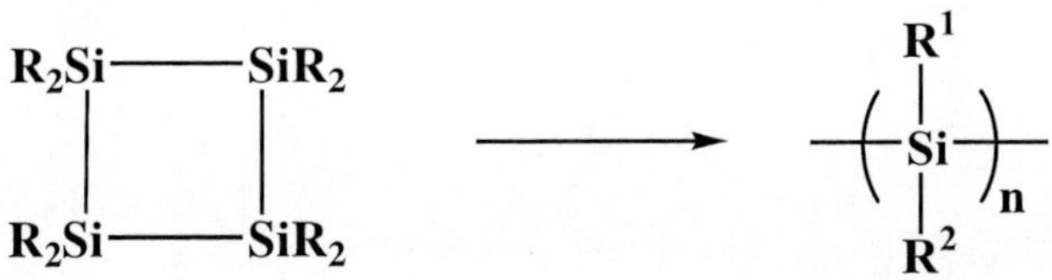

(1) (4)

The redistribution reaction of disilanes, $X_mR_{(6-m)}Si_2$(X=halogene, alkoxy), provide polymeric materials by the action of onium salts or alkoxides[17,18](eq. 5). However, the products have lost their linear structure but have highly branched or networked structures.

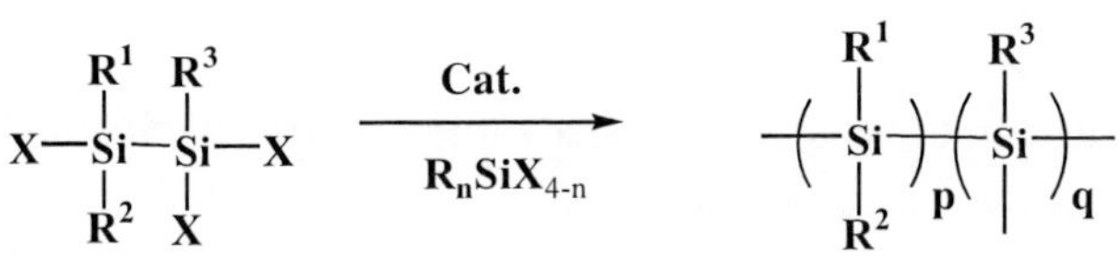

(5)

2.4. Polysilanes Having Functional Organic Groups

Organic groups have a decisive influence on the nature of polysilanes. The synthetic methods described in the above sections, unfortunately, provide only alkyl or aryl-substituted polysilanes with only a few exceptions. Most of the functional groups cannot survive under the drastic reaction conditions of using molten alkali metals. However, the functional groups that are located in a position further than gamma from the silicon atoms are able to resist the coupling reaction. Some polysilane homopolymers such as compounds 5-7 have been prepared by the Wurtz-type condensation reaction[19,20,21], .

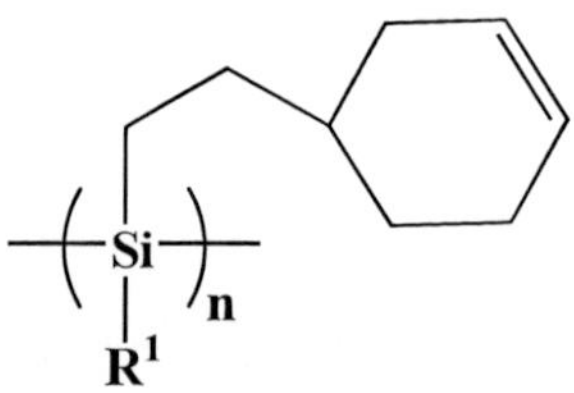

(5)

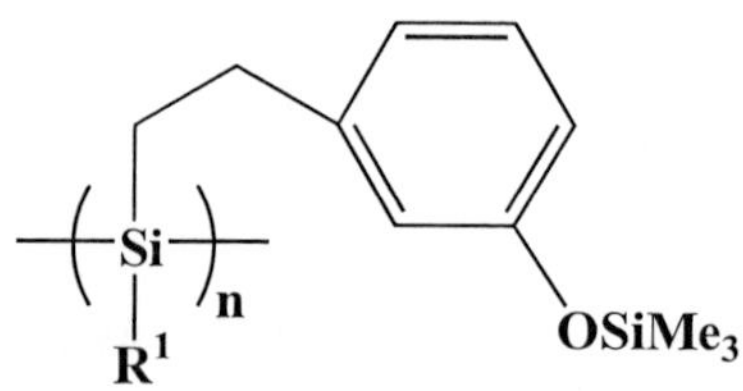

(6)

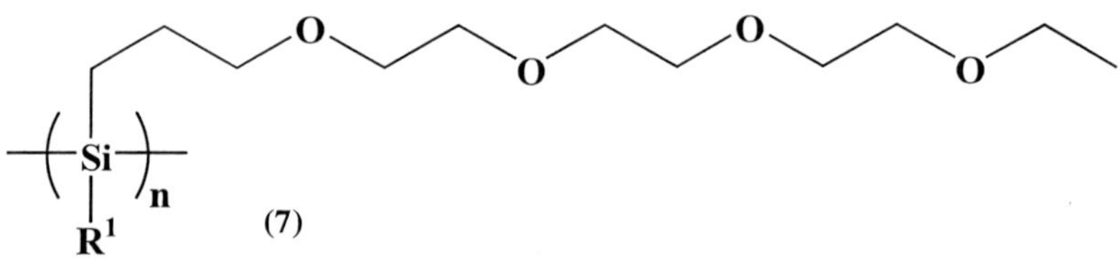

(7)

The polymer substitution reaction is widely adopted to obtain functional polysilanes while poly(methylphenylsilane)(PMPS) and poly(phenylsilane)(PMS) have been used for this purpose.

The treatment of PMPS with $HCl/AlCl_3$ and trifluoromethanesulfonic acid(TfOH) substituting the phenyl groups with chlorine atoms (eq. 6)[22] and trifluormethansulfonates (eq. 7)[23], respectively. The reaction of PMPS with $SnCl_4$-$ClCH_2OR$ introduces the chloromethyl-substituted phenyl groups (eq. 8)[24]. The chloro-, TfO- and chloromethyl-derivatives of PMPS were further converted to PMPS having various organic functional groups. The phenyl group of PMPS can also be utilized for arene-type complex formation (eq. 9), though the products were not stable enough for further studies.

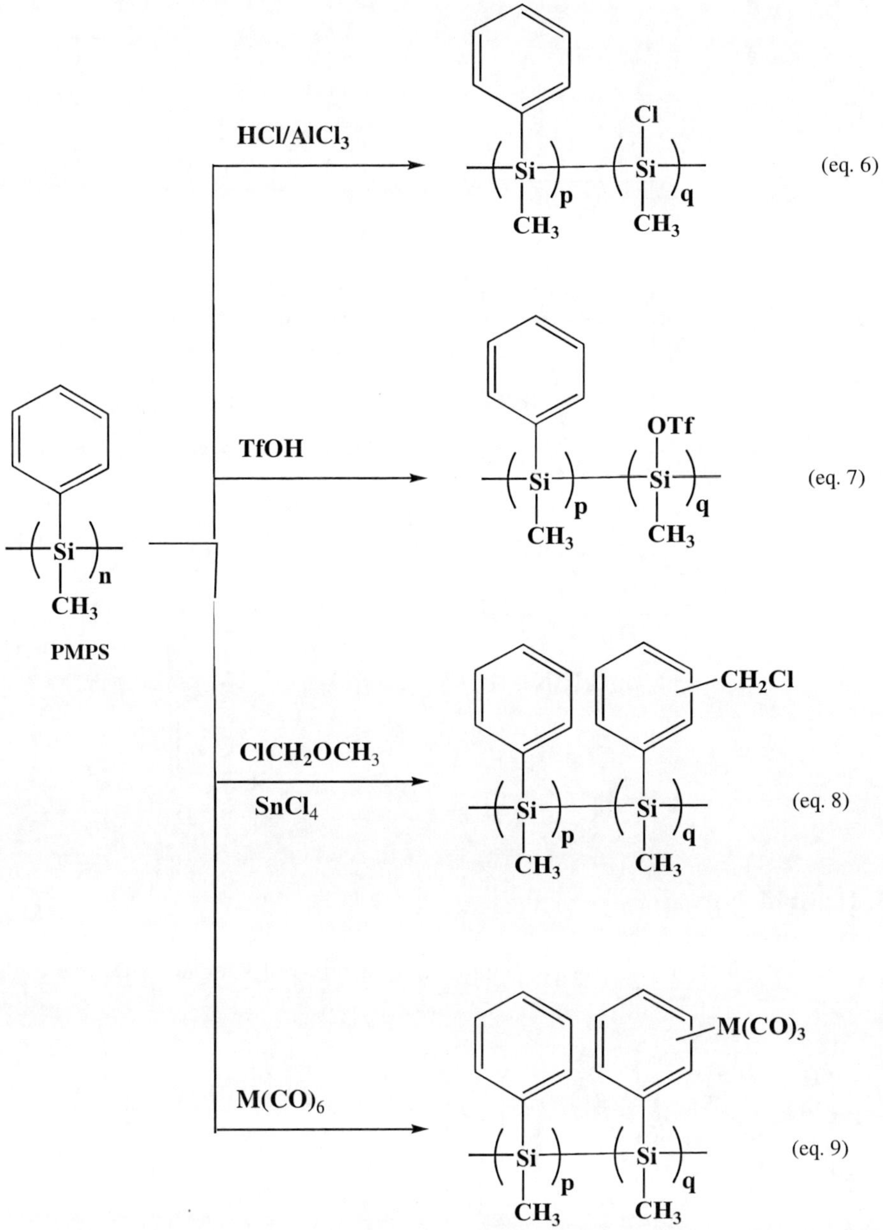

PMS has been used as ersatile starting materials for the production of various functional polysilanes via a chlorine-substituted reaction (eq. 10)[25] or hydrosilylation reaction either by H_2PtCl_6[26] or AIBN[12] (eq. 11).

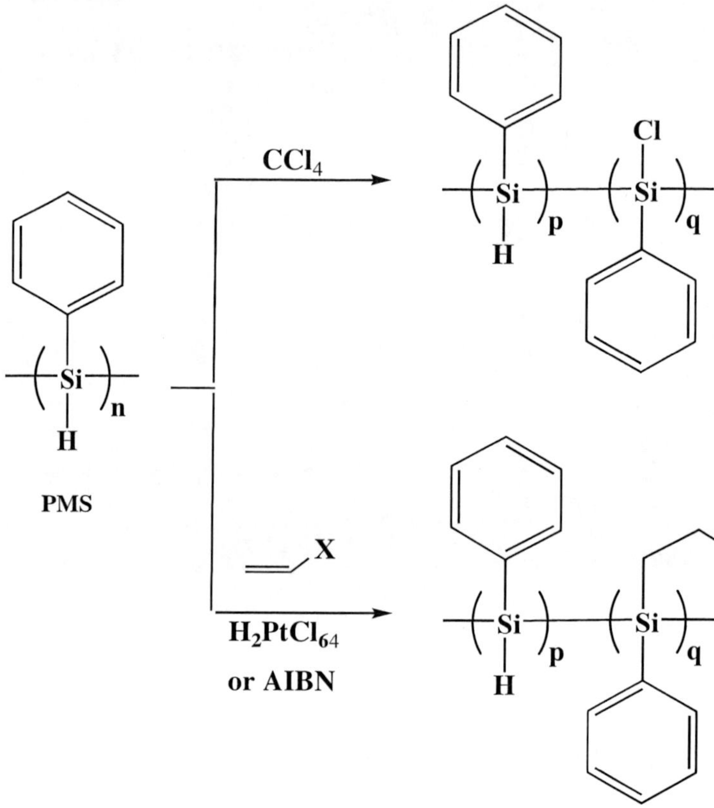

(eq. 10)

(eq. 11)

2.5. Hybrid Polymers

Block copolymers having polysilane and carbon-polymer parts (**8**) were also prepared.

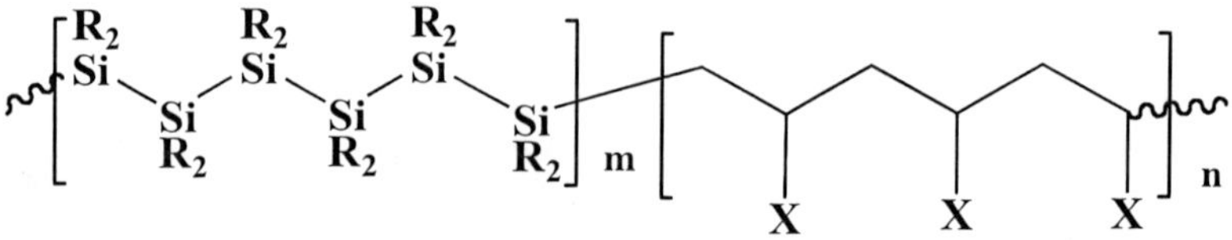

(8)

Halogen-end-caped polysilanes were obtained either from an unquenched mixture of an alkali condensation reaction or halogenative partial degradation of polysilanes. They were reacted with the living carbon polymer species (eq. 12)[27] and polyethylene (eq. 13)[28] to afford

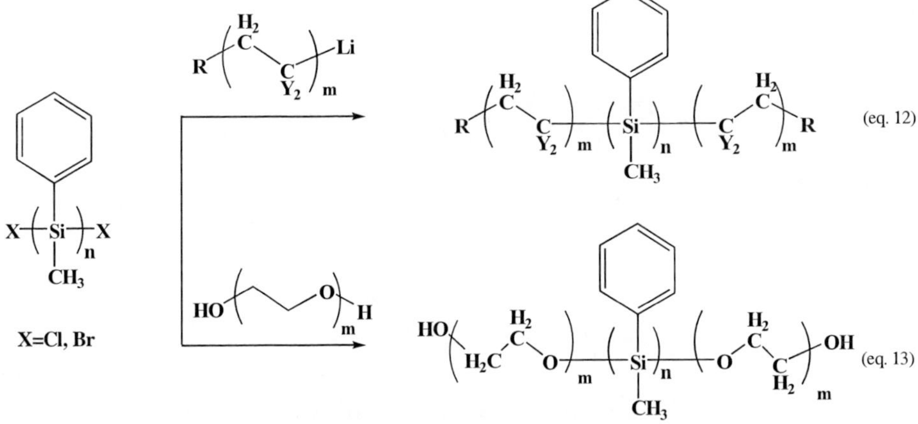

corresponding block copolymers. PMPS containing bromophenyl groups (eq. 13)[29] and chloromeththylphenyl-endo groups (eq. 14)[30] were subjected to an atom transfer polymerization reaction to give the corresponding grafted PMPS block polymers. Photolysis of PMPS produces the polysilanyl radical which initiates polymerization of many of the acrylic monomers to yield block polymers [eq. 15][31]. The unquenched products of the living polymerization reaction of Masked disilenes were used to initiate the polymerization of carbon monomers [eq.16][32].

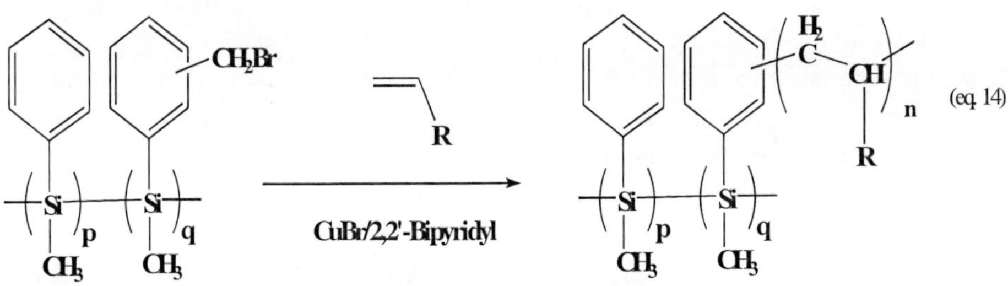

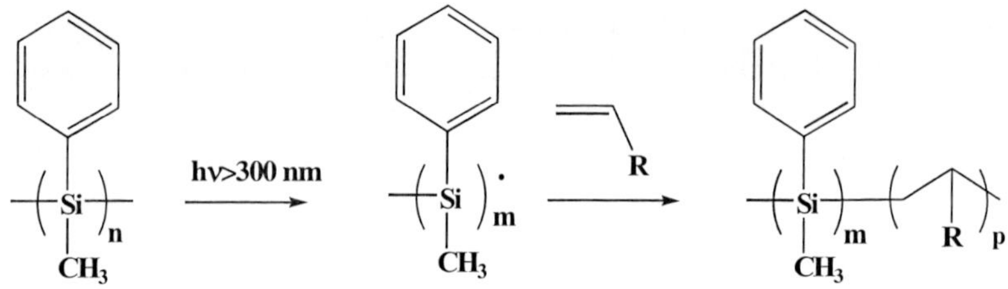

(eq. 15)

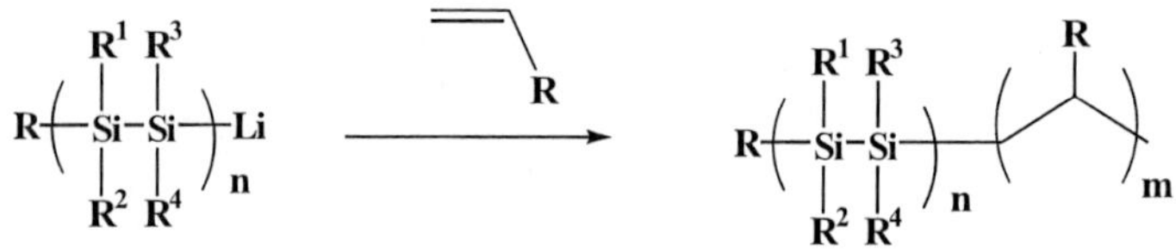

(Eq. 16)

3. Properties of Polysilanes

Homopolymers with dimethyl, diethyl, dipropyl-, and diaryl-substituents are insoluble crystalline solids. They decompose without melting upon heating. Most polysilanes baring larger organic groups than propyl groups are, in general, soluble and tractable. They show no melting points but show a glass transition temperature (T_g) from -70 °C to ~150 °C depending on the side groups[33].

3.1. Sigma Conjugation

The backbone of polysilanes solely consists of sp^3 silicon atoms. However, the chemical and physical properties of polysilanes resemble those of polyenes rather than polyolefiens.

A low ionization potential (~5.6 eV) is one of the typical characteristics of polysilanes[34], which indicates that the sigma electrons bond loosely to the silicon nuclei and, hence, delocalize over the backbone (sigma conjugation). This prominent delocalization of the sigma electrons, compared to those in carbon polymers, brings about similar electronic and optical properties to the pai-conjugated systems in polyenes. In accordance with the low ionization potential, polysilanes are known to form charge transfer complexes with the acceptors[35]. Furthermore, polysilanes become an electric conductor upon doping with oxidants such SbF_5, I_2, or $FeCl_3$., though neat polymers are insulating[36]. Moreover, hetero-atom substituted polysilanes show high conductivity (~10^{-2}S.cm^{-1}).

3.1.1. Electronic Spectra

Polysilanes shows strong absorption in UV region (300 - 410 nm) with absorbance >10^3/Si atom (figure 2).

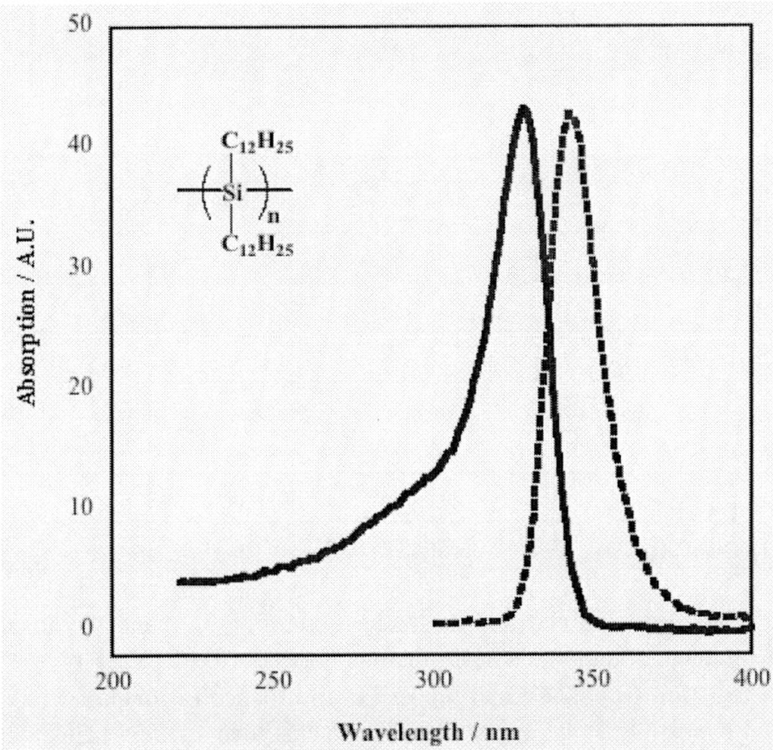

Figure 2. Absorption and emission spectra of didodecylsilane in THF.

The spectroscopic behavior of polysilanes is highly dependent on the organic pendants on the silicon atoms. In solution, many polysilanes with unbalanced alkyl groups (such as one long alkyl group and one methyl group) usually show λ_{max} of around 300 - 320 nm and the intensities are rather low (<5x10^3 /Si). Polysilanes with two long alkyl side groups at low temperature shows intense absorption (10^4/Si) of around 340 - 355 nm. Poly(arylalkylsilanes) absorb light at around 340 - 360 nm. And poly(diarylsilanes) display intense absorption (>10^4/Si) at around 400 nm. Thus, π-electrons of the aryl groups have some contribution on the sigma conjugation (σ–π conjugation). UV data of various polysilanes are summarized in table 1.

Table 1. Absorption and Fluorescence properties of various polysilanes [37, 38]

Polysilane	λ_{max}/nm	E_{max}/nm	Φ/%
$(Me\text{-}Si\text{-}Ph)_n$	341	353	0.08-0.25
$(Me\text{-}Si\text{-}C_6H_4OMe)_n$	344	362	0.2-0.5
$\{Me\text{-}Si\text{-}Np(b)\}_n$	350	437	
$(Me\text{-}Si\text{-}C_3H_7)_n$	306	341	0.76
$(Me\text{-}Si\text{-}C_6H_{13})_n$	318	337	0.61
$(Me\text{-}Si\text{-}C_{12}H_{25})_n$	309	337	0.67
$(C_6H_{13}\text{-}Si\text{-}C_6H_{13})_n$	317	342	0.42
$(Me\text{-}Si\text{-}{}^iPh)_m(Me\text{-}Si\text{-}{}^nPr)_m$	330	340	0.69
$(Me\text{-}Si\text{-}SiMe_3)_n$	312	335	0.026
$\{(p\text{-}BuC_6H_4)_2Si\}_n$	395	400	0.55
$\{(p\text{-}BuOC_6H_4)_2Si\}_n$	399	412	0.6

Recent studies have clarified the factor governing λ_{max} and the absorbance. Fujiki et al have found the linear relation between the a-value of the Mark-Hawkins equation $[\eta = kM^a]$ and the molar absorbance (ε) or half width of the absorption bands[39]. Polysilanes with chiral substituents take a helical structure with a very long persistence length and have unusually intense absorption ($\varepsilon > 10^5$). On the other hand, Tamao-synthesized oligosilanes with a well-defined conformation have demonstrated that the absorption maximum (λ_{max}) is related to the backbone conformation[40].

An early band calculation[41] of $(SiH_2)_n$ indicates that polysilanes are direct transition-type semiconductors with a band gap of 3-4 eV. A direct type band structure means a pronounced excitation (light absorption) as well as effective emissions. The calculations also indicate small values for the effective masses of electrons (0.15) and holes (0.11). These values are indicative of efficient σ-conjugation and high carrier transport properties.

3.2. Chromotropism of Polysilanes

The potential energy of polysilanes is dependent on the dihedral angle of the backbone and is sensitive to changes in the energetic circumstance. Recent theories[42] and experimental evidence[40] have clarified that the potential energy minima of most polysilanes are different from the carbon analogue and dependent on the ratio of the van der Waals radius of the organic groups to the Si-Si bond distance. Gauche and anti, familiar stable torsional angles in carbon compounds, are no more valid in polysilanes when the organic groups are larger than ethyl. The potential minima in these polysilanes are located at 40° (Cisoid, C), 90° (Ortho, O), 150° (Deviant, D), and 165° (Transoid,T)[42]. Studies on octasilane with CACAC conformation clearly indicate that no sigma-conjugation takes place through cisoid bonding[43].

The rotational barrier, however, seems small and thus intermediate torsional angles are possible by several factors such as crystal packing. It leads to the selectivity of the conformational change on the energy change of the circumstance. This resulted in a variety of chromotropic behaviors. In fact, polysilanes show thermo-, piezo-, electro-, solvato-, iono- and photo-chromisms..

When polysilanes, both in solid state and in solution, are cooled, the absorption maximum generally shift to longer wavelengths (red-shift), indicating that they take larger torsional angles[44]. In general, polysilanes display three types of thermochromic shifts upon cooling depending on the balance of the side groups; continuous red-shift, abrupt red-shift and intermediate type shifts (figure 3). However, careful studies revealed that the thermochromic behavior of polysilanes were more complicated than previously believed[45,46], which may reflect the small conformationa barrier of the Si-Si bonds.

Some polysilanes such as $(Bu_2Si)_n$ and $(Pen_2Si)_2$ in solid state shows a blue-shift upon cooling[47]. These polysilanes take a 7/3 helix coil as the extended form at low temperature, which corresponds to a dihedral angle of 154°. Heating the polysilanes loosens the coil resulting in the expanding of the dihedral angle. A larger dihedral angle implies longer wavelength absorption. Some poly(dialkylsilanes) also show a themochromism-like red-shift upon applying high pressure (piezochromism) [48].

Organic groups on the silicon atoms play an important role in the above mentioned chromotropic behavior through governing of the conformation. Poly(diarylsilanes) favor taking the conformation of a larger dihedral as their stable form. Polymers with unbalanced dialkyl groups take a more congested form.

Polysilanes with polar pendant groups display a red shift when interacting with polar solvents (solvatochromism)[21,49] or ionic species (ionochromism) through an enhancement of the bulky-size of the side groups, subsequently resulting in the arrangement of the main chain to a more extended form[50]. A spectral shift was induced by applying either an electric (electrochromism)[51] or magnetic field (magnetochromism)[52].

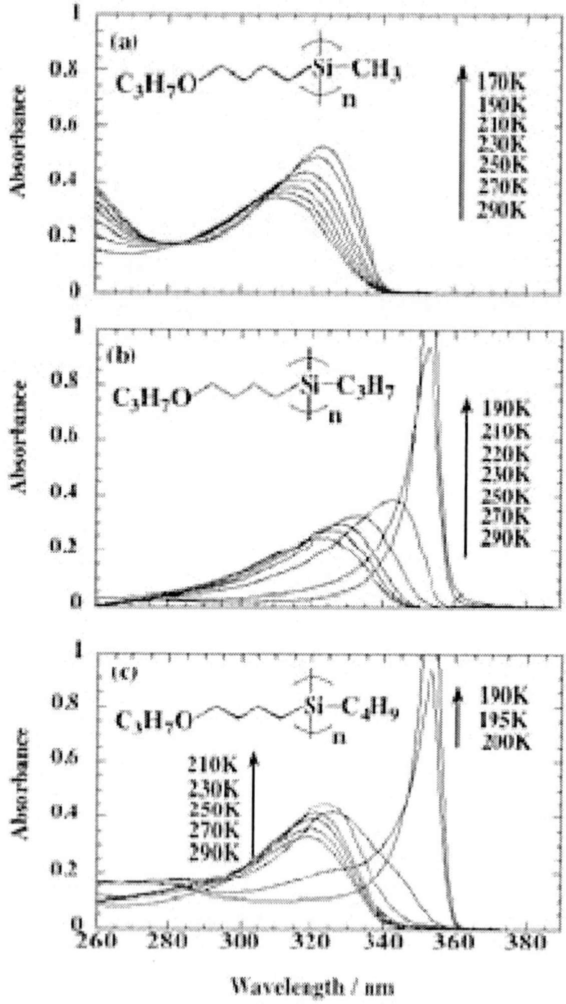

Figure 3. Three types of thermochromisms of poly(dialkylsilane)s.

3.3. Photochemistry of Polysilanes

Photolysis of trisilanes yields silyl radicals, silylenes, and silenes as the reactive intermediates (Scheme 3)[53]. Photolysis of high molecular weight polysilane undergoes similar processes as the above, i.e., in the presence of trapping agents, polysilanes degrade *via* a homolytic process (path a) as the main course and silylene extrusion (path b) as the minor process. The radical species undergo a secondary process to yield Si=C intermediates (path c) which are also detected. Photolysis of aryl-substituted polysilanes in the solid state brings about a more complicated situation; both chain scission and cross-linking take place, though in most cases, degradation is still the main process. In solution, the main chain scission is almost the sole process resulting in low molecular weight polymers.

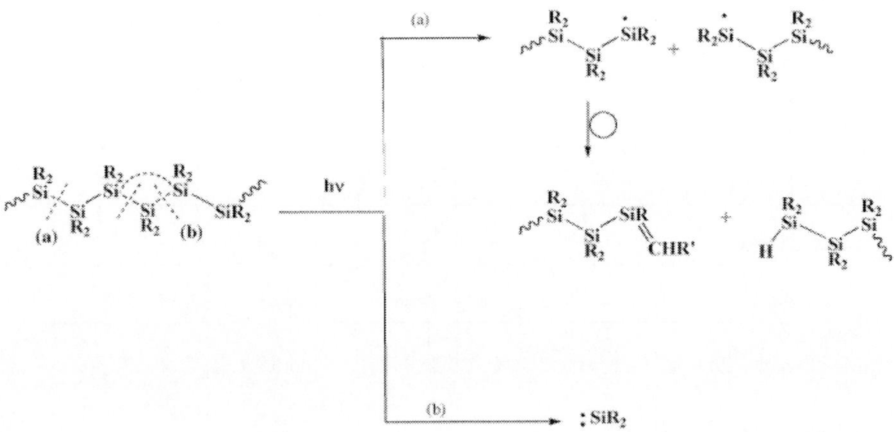

Scheme 3.

3.4. Carrier Transport of Polysilanes

Polysilanes, like conjugated carbon polymers, are insulators without doping. Photo-irradiation on solid polysilane films, however, induces photocurrents (photoconductivity)[54]. The conductivity of polysilane is very high compared with organic polymers (~$10^{-4}cm^2V^{-1}s^{-1}$). Some corrective values are listed in table 2[55]. Carrier transport is explained, as in carbon polymers, by a hopping model where carriers migrate the sigma-conjugated domains by hopping (figure 4)[54].

Therefore, alignment of the σ-conjugated domain is important for polysilanes to achieve higher transport properties.

Table 2. Ionization potential and hole drift mobility of polysilanes [53]

Polysilane	$Mw/10^{-3}$	Ip/eV	$\mu/(cm^2/Vsec)$
(Me-Si-Me)		5.73	
$(Bu)_2Si$		5.7	5.3
$(Pen)_2Si$		5.9	
$(Hex)_2Si$		5.8	4.5
$(p\text{-}Bu\text{-}O\text{-}Ph)_2Si$		5.5	4
Me-Si-Ph	69	5.6	$3.5x10^{-4}$
Me-Si-p-Tolyl	9.5	5.45	$32.3x10^{-5}$
$Me\text{-}Si\text{-}C_6H_4\text{-}C_6H_5$	77	5.59	$2.5x10^{-5}$
Me-Si-Pr	71	5.66	$1.4x10^{-4}$
$Me\text{-}Si\text{-}CH_2CH_2Ph$	210	5.64	
Me-Si-c-Hex	120	5.59	
Pr-Si-Hex	150	5.62	
$Me\text{-}Si\text{-}SiMe_2Ph$	180	5.85	
$Me\text{-}Si\text{-}SiMe_2(p\text{-}Tolyl)$	19	5.52	
$Me\text{-}Si\text{-}SiMe_2(o\text{-}Tolyl)$	8.4	5.57	
$(Me\text{-}Si\text{-}Me)_n(Me\text{-}Si\text{-}Ph)_m$	71	5.67	$3.5x10^{-4}$
$(Me\text{-}Si\text{-}Me)_n(Me\text{-}Si\text{-}CH_2CH_2Ph)_m$	48	5.84	$4.7x10^{-7}$
$(Me\text{-}Si\text{-}c\text{-}Hex)_n(Me\text{-}Si\text{-}Ph)_m$	3.7	5.79	$1.3x10^{-5}$
$Me\text{-}Si\text{-}c\text{-}Hex)n(Me\text{-}Si\text{-}p\text{-}Tolyl)_m$	28	5.57	
$Me\text{-}Si\text{-}Hex)n(Me\text{-}Si\text{-}Ph)_m$	2.9	5.77	
$(Me\text{-}Si\text{-}Ph)_n(Me\text{-}Si\text{-}p\text{-}Tolyl)_m$	2.9	5.79	$4.5x10^{-5}$
$(Me_2Si\text{-}SiMePh)_n$	900	5.64	$6.0x10^{-}$

For this purpose, studies of the ordering of polysilane chains have been reported including thermal treatment[56], friction treatment[57], a LB method[58], electric field treatment[59], magnetic field treatment[52], and rubbing[60] .

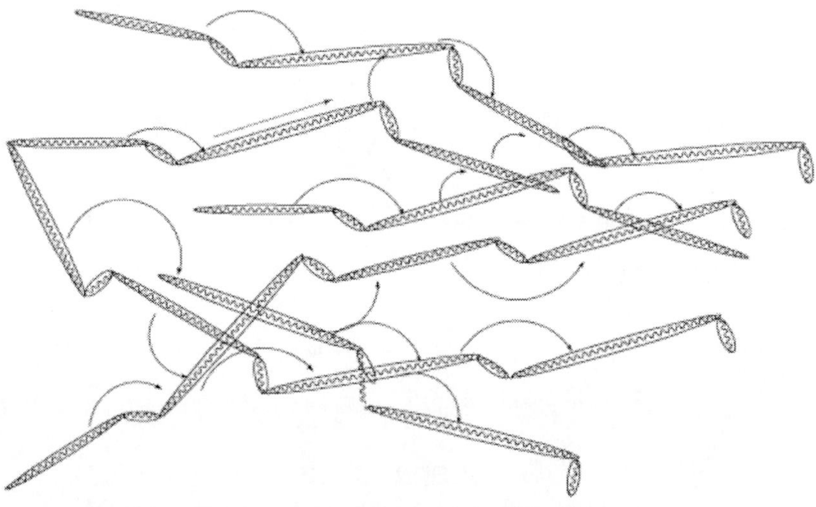

Figure 4. Hopping model of carrier transports in polysilanes.

3.3. Some Possible Technological Application of Polysilanes

Polysilanes were investigated as new photo-resists[61] because of their strong absorption in the UV region and efficient photo-scission process. However, the requirement of a high dose is a disadvantage for industrial applications. Interestingly, polysilanes were utilized as the photoresist for the production of a color filter of LCD by a sol-gel processutilizing efficient photo-scission process and the hydrophylicity of the resulting product[62].

Electrophotographic applications have been extensively studied by the virtue of their high carrier transport characteristics[63]. Transparent polymers with high carrier mobility and high fluorescence quantum yields can be used as emitting layer materials for organic LEDs. Polysilane LEDs emitted almost pure UV light (figure 5). A low emission efficiency (<0.2%) was mainly attributed to the low electron mobility of polysilanes[64,65]. On the other hand, by virtue of their high hole mobility, polysilanes are studied as the hole tansport layer of LEDs. However, polysilanes are only used for industrial production of silicon-carbide fiber which has higher tensile strength at high temperatures and are more resistive to oxidation that carbon fibers[66].

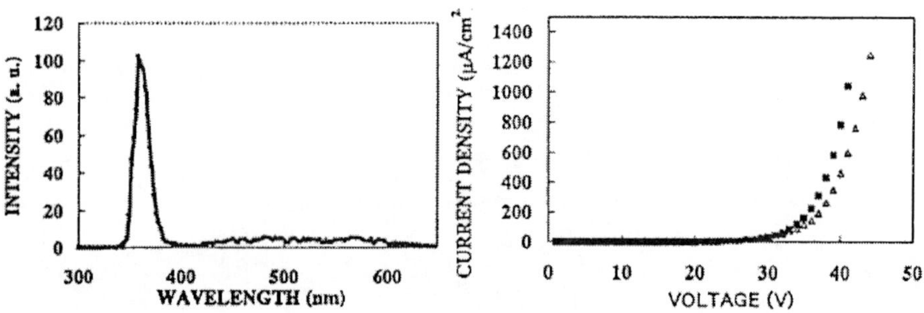

Figure 5. Emission spectrum (left) and I-V characteristic of PMPS LED at room temperature.

References

[1] Mazdyasni, K.S; West, R.; David, L.D. J. Am. Ceram. Soc. 1978, 61, 504.West, R.; David, L. D.; Djurovich, P. L.; Stearley, K. L.; Srinivasan, K. S.; Yu, H. J. *Am. Chem. Soc.* 1981, 103, 7352.

[2] McLiesh, T. C. B.; Jones, R. G.; Holder, S. J. *Macromolecules.* 2002, 35, 548.

[3] Oka,K.; Fukudome, M.; Dohmaru, T.; West, R. *Electrophotography.* 1995,34, 386.

[4] Zeigler, J. M. *Polym. Prepr.* 1986, 27, 109.

[5] Miller, R.D.; Thompson, D.; Sooriyakumaran, R. Fickes, G. J. *Polym. Sci. Part A: Polym Chem.* 1991, 29,813..

[6] Matyjaszewski, K.; Grezta, D.; Hrkash, J. S.; Kim, H. K. *Macromolecules.* 1995, 28, 59.

[7] Fujino, M,; Isaka, H; *J.* Chem. Soc., Chem. Commun. 1989, 466.

[8] Shono, T.; Kashimura, S.; Ishifune, M.; Nishida, R. *J. Chem. Soc. Chem. Commun.* 1990, 1160.

[9] Matyjaszewski, K. *Polym. Prepr.* 1987, 28, 224.

[10] Aitken, C.;Harrod, J. F .; Samuel, E. J. *Organomet. Chem.* 1985, 279, C11.
[11] Woo, H. G.; Walzter, J. F.; Tilley, T. D. *J. Am. Chem. Soc.* 1992, 114, 7074.
[12] Hsiao, Y. L.: Waymouth, R. M. *J. Am. Chem. Soc.* 1994, 116, 9779
[13] Sakamoto, K.; Obata, H.; Hirata, H.; Nakajima, M.; Sakurai, H, *J. Am. Chem. Soc.* 1989, 111, 7641.
[14] Crypryk, M.; Chrusciel, J.; Fosson, E.; Matyjaszewski, K. *Makromol. Symp.* 1993, 73, 167.
[15] Koe, J. R.; Powell, D. P.: Buffy, J. J.; Hayase, S.; West, R. *Angew. Che. Int. Ed. Engl.* 1998, 10, 37
[16] Suzuki, M.; Kotani, J.; Gyobu, S.; Kaneko, T.; Saegusa, T. *Macromolecules.* 1994, 27, 2360.
[17] Baney, R. H.; Gaul, Jr. J. H.;Hilty, T. K. *Organometallics,* 1983, 2, 859.
[18] Watanabe, H.; Higuchi, K.; Kobayashi, M.; Kitahama, T.; Nagai, Y. *J. Chem. Soc. Chem. Commun.* 1977, 704.
[19] Stueger, H; West, R. *Macromolecules.* 1985, 18, 2349.
[20] Horiguchi, R.; Onishi, Y.; Hayase, S. *Macromolecules.* 1988, 21, 304.
[21] OKa, K.; Fujiue, N.; Dohmaru, T.; Yuan, C-H.; West, R. *J. Am. Chem. Soc.* 1997, 119, 4074.
[22] West, R. J. *Organomet. Chem.* 1986, 300. 327.
[23] Matyjaszewski, K;. Chen, Y. L.; Kim, H. K. ACS Symp. Ser. 1988, 360, 78.
[24] Ban, H.; Sukegawa, K.; Tagawa, S. *Macromolecules.* 1987, 20, 1775.
[25] Banovetz, J. P.; Hsiaso, Y. L.; Waymouth, R. M. *J. Am. Chem. Soc.*, 1993, 115, 2540.
[26] Qui, H.; Du, J. J. Polym. Sci. Part A: Polym. Chem. Ed. 1989, 27, 2861.
[27] Demoustier-Champagne, S.; de Mashie, A. F.; Devaux, J.; Fayt, R.; Teysie, P. *J. Polym. Chem.* 1993, 31, 2009.
[28] Holder, S.; Hioms, R. C.; Sommerdijk, N. A. J. M.; Wiiliams, S. J.; Jonews, R. G.; Notle, R. J. M. *J. Chem. Soc. Chem. Commun.* 1998, 1445.
[29] Lussen, L.; Cordina, G. P. G.; Jones, R. G. *Eur. Polym. J.* 1998, 34, 1829.
[30] Jones, R. G.; Benfield, R. E.; Swain, A. C.; Webb, S. J.; WENT, M. J. *POLYMER.* 1995, 36, 393.
[31] West, R.; *J. Radn. Curing.* 1986, 13, 35-40.
[32] Sakamoto, K.; Obata, K.; Hirata, H.; Nakajima, M.; Sakurai, H. *J. Am. Chem. Soc.* 1989, 111, 7641.
[33] Demoustier-Champagne, S.; Devaux, J. in Silicon-Containing Polymers, Jones, R. G.; Ando, W.; Chojnowski, J. *"Thermal Properties and Phase Behaviour of Polysilanes",* Dordrecht, Kluwer Academic Publishers,, 2000, 553-574.

[34] (a)Cott, P. M. *J. Polym. Dci. Part B. Polym. Phys.* 1994, 32, 771.(b)Yonezawa, Y.; Dohmaru, T.; Oka, K. Unpublished studies.
[35] Traven, V. F.; West, R. J. *Am. Chem. Soc.*, 1973, 95, 6824,.
[36] Fukushima, M.; Hamada, Y.; Tabei, E.; Aramata, M.; Mori, S. *Synth. Metals*. 1998, 94, 239.
[37] Miller, R. D.; Michl, J. *Chem. Rev.* 1989, 89, 1359-1410.
[38] L. H. Harrah, J. M. Zeigler, in *"Photophysics of Polymers, ACS Symp. Ser.* 358, Hoyle, C. H. ; Torkelson, J. M. :1987, P482
[39] Fujiki, M. J. *Am. Chem. Soc.* 1996, 118, 7425
[40] Forgarty, H. A.; Ottoson, C. H.; Michl, J. J. *Mol. Struct.(theochem)*. 2000, 104, 3418.
[41] Takeda, K.; Matsumoto, N.; Fukuchi, M. *Phys. Rev.* 1984, B30, 5871.
[42] West, R. J. *Organomet. Chem.* 2003, 685, 6.
[43] Tsuji, H.; Terada, M.; Toshimitsu,, A.; Tamao, K. J. *Am. Chem. Soc.* 2003, 125, 7468.
[44] Schilling, F. A.; Bovey, F. A.; Lovinger, A. J.; Zeiglere, J. M. *ACS Adv. Chem. Ser.* 1990, 224, 341-378.
[45] Bukalov, S. S.; Leites, L. A.; West, R. Asuke, T. *Macromolecules*. 1993, 26, 3937.
[46] Chunwachirasiri, W.; West, R.; WInokur, M. J. *Macromolecules*. 2000, 33, 9720.
[47] Schilling, F. A.; Lovinger, A. J.; Zeiglere, J. M. ; Davis, D. D.; Bovey, F. A. *Macromolecules*. 1989, 22, 3055.
[48] Moller, M.; Frey, H.; Sheiko, S. *Colloid Polym. Sci*. 1993, 271, 554.
[49] Oka, K.; Fujiue, N.; Nakanishi, S.; Takata, T.; Dohmaru, T.; West, R. J. *Organom,et. Chem.* 2000, 611, 45-51.
[50] (a)Yuan, C. H.; West, R. Chem. Commun. 1997, 1825. (b)Oka, K.; Fujiue, N.; Nakanishi, S.; Takata, T.; Dohmaru, T.; Yuan, C. H.; West, R. *Chem. Lett.* 1997, 253-254.
[51] Fujino, M.; Hisaki, T.; Matsumoto, N. *Macromolecules*. 1995, 28, 5017.
[52] Nakayama, Y.; Nakagawa, H.; Oka, K.; Naito, H.; Mogi, I.; Watanabe, L. Dohmaru, T. *Solid State Commun.* 1998, 109, 45-49.
[53] Trefonas, P. T. III; West, R.; Miller, R. D. *J. Am. Chem. Soc.* 1985, 107, 2737.
[54] Abkowitz, M. A.; Stolka, M.; Weagley, R. J.; McGrane, K. M.; Knier, E. F. *ACS Adv. Chem. Ser.* 1990, 224, 467-503.
[55] Dohmaru, T.; Oka, K.; Yajima, T.; Miyamoto, M.; Nakayama, Y,; Kawamurea, T. ; West, R. *Phil. Mag.* 1995, B71, 1069-1081.

[56] Nakayama, Y.; Hirooka, K.; Oka, K. ; Dohmaru, T. *Solid State Commun.* 1998, 109, 45-49.

[57] Tanigaki, N.; K.; Yase, Kaito, A. *Thin Solid Films,* 1996, 273, 263.

[58] Yoshida, M.; Seki, T. Nakanishi, F. Sakamoto, A, Sakurai, H. *Chem. Commun.* 1996, 1381.

[59] Nakayama, Miyamot, M.; Akita, S.: Oka, K. ; Dohmaru, T.; West, R. *Jpn. J. Appl. Phys.* 31995, 4, L57.

[60] Tanigaki, N.; Yase, K.; Ueno, K. *Polymer.* 1995, 36, 2477

[61] Miller, R. D. *ACS Adv. Chem. Ser.* 1990, 224, 413-458.

[62] Yokoyama, M.; Kawabata, M.; Sumiyoshi, I.; Yokoyama, M. *Chem. Lett.* 1994, 1119.

[63] Stolka, M. Yuh, H. J.; McGrane, K.; Pai, D. M. *J. Polym. Sci. part A. Polym. Chem.* 1987, 25, 823.

[64] (a)Suzuki, H.; *Advan, Mater.* 1996, 8, 657. (b) Xu, Y.; Fujino, T,; Naito, H.; Oka, K.; Dohmaru, T. *Chem. Lett.*, 1998, 299-300.

[65] Yajima, S.; Hayashi, J.; Ohmori, M. *Chem. Lett.*, 1975, 931.

[66] Xu, Y.; Fujino, T.; Naito, H.; Oka, K.; Dohmaru, T. *Chem. Lett.* 1998, 299-300.

INDEX

#

3,4-ethylenedioxythiophene, 113

A

AAP, 27
abdomen, 8, 12, 13
absorption, 34, 50, 53, 65, 66, 68, 69, 70, 71, 77, 99, 104, 105, 106, 107, 108, 115, 116, 117, 118, 128, 132, 181, 182, 193, 194, 195, 199
absorption coefficient, 116
absorption spectra, 65, 66, 68, 105, 108
absorption spectroscopy, 107
acceptor, 94
acceptors, 192
accuracy, 151
acetaldehyde, 146
acetate, 5, 20, 183
acetic acid, 101
acetonitrile, 76, 97, 131
acetylcholine, 2, 5, 6, 8, 30
acetylene, 93
ACF, 169
achievement, 19
acid, 4, 76, 83, 84, 101, 131, 139, 140, 141, 142, 143, 144, 145, 146, 151, 154, 155, 157, 158, 164, 165, 167, 188
acidity, 76
acrylate, 138, 140, 145, 146
ACS, 201, 202, 203
activation, 61, 165
activators, 3
active site, 61
actuators, 77
additives, 43, 53, 90, 166, 183
adducts, 125
adhesion, 64, 165
adhesives, 137, 138, 139, 147
adsorption, 156, 171
aerobic, 11
AFC, 83, 84
AFM, 168, 171, 173, 174
Ag, 6, 38, 40
age, 137
agent, 5, 166
agents, 36, 153, 197
aggregation, 167
agricultural, 54
AIBN, 146, 189
aid, 137
air, 49, 50, 53, 62, 84, 131, 145
albumin, 151
alcohol, 36, 92, 141
alcohols, 131
aldehydes, 131
alkali, 37, 182, 185, 188, 191
alkaline, 77, 80, 84
alkane, 172
alkenes, 89, 90, 91, 92, 95, 102, 128, 130

alkynes, 124, 125, 126, 127, 128, 130, 131, 133
all-electric, 40
alloys, 43, 80, 87
alternative, 86, 133, 151, 157
aluminum, 36
amine, 26
amines, 131, 141, 154
amino, 151, 154
amino acid, 151, 154
ammonia, vii, 24, 25, 27, 49
ammonium, 9, 10, 24, 25
ammonium chloride, 24, 25
Amsterdam, 29, 86
annealing, 109, 110, 112
anode, 43, 46, 76, 81, 85, 112, 113
anthracene, 118
antibody, 3, 4, 150, 157
antibonding, 60
anticancer, 152, 153, 156
anticancer drug, 152, 153, 156
antigen, 150
antitumor, 156
application, 17, 22, 29, 34, 42, 45, 51, 53, 62, 81, 133, 156, 163, 169, 170, 171
aqueous solution, 1, 21, 22, 76, 77, 84, 97, 155, 167
aqueous solutions, 1, 21, 22, 155
aromatic hydrocarbons, 115
aromatic rings, 89, 90, 93
ascorbic acid, 166
atmosphere, 61, 69, 77, 107, 134, 145, 146
atoms, 38, 117, 133, 134, 169, 183, 185, 188, 192, 193, 195
Au nanoparticles, 170
availability, 53, 182, 187
awareness, vii, 137

B

baking, 140, 141, 142, 143, 144, 146
band gap, 52, 53, 65, 67, 194
barrier, 12, 168, 172, 195
barriers, 159
base pair, 174
basicity, 76
batteries, 34, 38, 40, 42, 43, 44, 45, 46, 47, 79, 80, 86, 87
battery, 39, 41, 42, 45, 46, 75, 77, 78, 79, 80, 81, 82
beams, 65
behavior, 1, 193, 195
benign, 89, 102
benzene, 90, 91, 92, 93, 94, 95, 97, 98, 100, 107, 135
binding, 160
biocompatibility, 156, 164
biocompatible, 152
biodegradability, 156
biodegradable, 152
biomass, 54, 86
biomolecule, 4
biomolecules, 1, 2, 4, 28
biosensors, 1, 2, 3, 4, 8, 29, 164
bisphenol, 138
block polymers, 191
blocks, 155
blood, 1, 2, 12, 29, 150, 155, 156
blood stream, 1, 29, 150, 156
boiling, 36, 37, 76
bonding, 60, 91, 165, 167, 169, 195
bonds, 50, 117, 125, 128, 145, 181, 185, 195
borate, 7
brain, 1, 2, 3, 8, 9, 11, 12, 13, 28, 29, 30
brain chemistry, 2
branched polymers, 151
bromine, 115
buffer, 6, 7, 9, 10, 12, 174
buildings, 81
bulk materials, 164
burn, 76
burns, 76, 86

C

Ca^{2+}, 32, 38
calibration, 6, 7, 9
calorimetry, 107
Canada, 85
cancer, 149, 156

candidates, 45, 61, 100
capillary, 30, 97
caprolactone, 155
carbon, 33, 43, 46, 50, 76, 85, 89, 100, 102, 103, 117, 124, 126, 128, 130, 132, 169, 171, 182, 190, 191, 192, 195, 197, 200
carbon atoms, 169
carbon dioxide, 50, 76, 132
carbon materials, 43
carbon monoxide, 126
carbon paper, 85
carboxylic acids, 132
carrier, 6, 7, 9, 10, 12, 41, 150, 159, 181, 194, 199
catalase, 6
catalysis, 49, 164
catalyst, vii, 35, 49, 56, 58, 60, 61, 65, 66, 67, 68, 85, 124, 125, 126, 141, 183
catalytic activity, 85, 131
catalytic system, 124
cathode, 43, 45, 46, 76, 81, 85, 112, 114
cation, 100, 102, 104, 105, 106, 107, 108, 109
cavities, 61, 72
cell, 38, 45, 46, 47, 75, 77, 78, 79, 81, 82, 83, 85, 86, 154, 158, 159, 160
cell surface, 159
cellulose, 137
cellulose derivatives, 137
central nervous system, 8
ceramic, 42, 43, 44, 45, 46
ceramics, 41, 42, 44
CH4, 56, 58
chain scission, 197
chemical energy, 50, 53, 54, 56, 76, 82
chemical engineering, 2
chemical properties, 33, 36
chemical reactions, 15, 90, 113, 115, 117
chemical sensing, 30
chemical stability, 53, 85, 138
chemicals, 49, 138, 163, 164, 166
chemiluminescence, 30
chemotherapy, 156
chitosan, 151
chloride, 13, 14, 25
chlorine, 188, 189
chromium, 131
circulation, vii, 150, 156
cis, 91
classical, 3
classification, 83
clean energy, 75
cleaning, 63, 64
cleavage, 100, 104, 111, 112, 128, 141, 146
climate change, 49
clusters, 57
CMC, 155
Co, 80, 81, 85, 86, 122
CO2, 50, 53, 54, 56, 58, 62, 67, 70, 72, 82, 84, 86, 132
coatings, 36, 45, 137, 138, 139
cobalt, 126
cofactors, 3
coil, 5, 6, 195
colloids, 164
colors, 164
combustion, 76
communication, 34, 108
community, vii
compatibility, 163, 177
compensation, 183
components, 18, 85, 155, 172
composites, 35, 45, 152
composition, 38, 42, 71, 80
compounds, 50, 53, 56, 61, 62, 86, 89, 90, 94, 95, 100, 115, 116, 123, 124, 125, 126, 127, 128, 130, 133, 141, 165, 182, 186, 187, 188, 195
computing, 170
concentration, 1, 2, 10, 11, 12, 14, 22, 40, 41, 96, 155, 183
condensation, 133, 182, 183, 185, 188, 191
conductance, 168, 171
conduction, 34, 38, 39, 41, 52, 56
conductive, 40, 43, 46, 163, 165, 167, 168, 169
conductivity, 38, 39, 40, 41, 42, 44, 45, 76, 85, 168, 173, 174, 177, 181, 192, 197
conductor, 34, 163, 169, 192
configuration, 4
conjugated dienes, 128, 129

conjugation, 134, 151, 181, 192, 193, 194, 195
consciousness, 85
construction, 3, 17, 94, 97, 115, 133, 170
consumption, 49, 50
contact time, 18
contamination, 18
continuity, 46
contracts, 76
control, 69, 90, 131, 154, 156, 174, 181
conversion, 6, 34, 50, 53, 54, 72, 82, 98, 99, 117, 145
cooling, 16, 195
copolymer, 143
copolymerization, 143, 157
copolymers, 143, 144, 151, 155, 156, 182, 190, 191
copper, 165, 172
core-shell, 45, 164
corona, 155
cost-effective, 174
coupling, 1, 3, 9, 107, 139, 168, 182, 183, 185, 188
covalent, 165
CPG, 5
CRC, 177
crosslinking, 5, 140, 141, 142, 144, 197
crosslinking reactions, 140, 142
cross-sectional, 22, 24, 71
crystal growth, 35
crystal lattice, 77
crystal structure, 38, 39, 41
crystal structures, 38, 39
crystalline, 36, 39, 40, 41, 42, 157, 192
crystalline solids, 192
crystallization, 40, 42
crystals, 4, 40, 42, 71
curing, 139, 140, 145, 146, 147
curing process, 139
cycles, 44, 80, 174
cycling, 45, 86
cyclohexane, 92, 116
cyclohexyl, 143
cysteine, 12, 154
cysteine residues, 154
cytosol, 151, 158, 159

D

DCA, 104, 106, 107, 108
decay, 57, 106, 108, 111, 112, 117, 118
decomposition, 50, 60, 61, 65, 66, 67, 69, 113, 146
defects, 38
degenerate, 103, 104, 106, 107, 109
degradation, 50, 62, 63, 138, 139, 146, 147, 191, 197
dehydration, 156
dehydrogenase, 10
delivery, 149, 150, 151, 154, 156, 157, 158, 159, 185
delocalization, 192
dendrimers, 149, 151, 152, 153, 154, 155
density, 76, 107
density functional theory, 107
deposition, 54, 64, 69, 70, 72, 165, 175
derivatives, 90, 92, 103, 116, 118, 126, 134, 137, 139, 143, 145, 183, 188
desorption, 77
destruction, vii
detection, 3, 8, 9, 10, 13, 17, 18, 27, 28, 29, 30, 164, 173, 174, 176
detergents, 155
deviation, 80
DFT, 107, 111
dialysis, 7, 10, 13, 14
diffraction, 42
diffuse reflectance, 68
diffusion, 15, 16, 19, 25, 85, 99, 154, 185
diffusion time, 15, 16, 25
dimeric, 102
dimerization, 139, 186
diodes, 109, 112
dioxins, 62
dipole, 76, 92
dipole moment, 76, 92
Discovery, 160
diseases, 149, 174
disorder, 174
dispersion, 21, 85, 174, 183

dissociation, 155
distillation, 75
distribution, 22, 135, 150, 151, 152, 183
disulfide, 129, 134, 154
DNA, 3, 159, 163, 170, 171, 173, 174, 176, 179
donors, 100
dopant, 114
doped, 65, 67
doping, 35, 65, 114, 181, 192, 197
double bonds, 128
drug carriers, 150, 152, 156
drug delivery, 149, 151, 153, 154, 155, 156, 157
drug delivery systems, 153, 157
drug release, 151, 156, 157, 160
drugs, 149, 150, 151, 154, 156, 160
durability, 18, 29, 86, 109, 112, 113, 114, 115, 137
dyes, 3, 18

E

earth, 34, 53, 65, 67, 76, 80, 123, 132
ecology, 75
EDOT, 113
elastomers, 137
electric current, 44
electric energy, 76, 82
electric field, 68, 151, 199
electricity, 81, 82
Electroanalysis, 30, 31, 178
electrochemistry, 86
electrodeposition, 165
electrodes, 3, 4, 13, 44, 47, 76, 78, 81, 85, 174, 175, 183
electroluminescence, 112
electrolysis, 50, 78, 81, 86
electrolyte, 34, 38, 42, 43, 44, 46, 78, 83, 85, 86
electrolytes, 38, 40, 41, 42, 43, 44, 45, 46, 47, 76
electromotive force, 82
electron, 34, 37, 52, 56, 60, 67, 68, 89, 90, 93, 94, 100, 103, 104, 106, 112, 130, 131, 132, 168, 172, 200
electron paramagnetic resonance, 104
electronic structure, 52, 68, 71, 104, 105
electrons, 52, 54, 56, 61, 62, 67, 181, 192, 193, 194
electron-transfer, 100, 132
electroplating, 165
electroreduction, 165
electrostatic force, 159
emission, 37, 92, 106, 109, 111, 112, 113, 114, 115, 182, 193, 199
encapsulated, 154
encapsulation, 152, 153, 154
endocytosis, 154, 158, 159
energy, 34, 38, 39, 42, 45, 46, 47, 49, 50, 52, 53, 54, 56, 62, 63, 65, 67, 72, 76, 77, 79, 82, 86, 96, 107, 112, 113, 116, 118, 124, 137, 138, 139, 195
energy density, 38, 42, 45, 46, 47, 79
energy recovery, 137
engines, 82
entropy, 76, 82
environment, vii, viii, 49, 67, 68, 151, 154, 158
environmental awareness, 137
environmental contaminants, 53
environmental issues, 72
environmental protection, 51
enzymes, 3, 5, 14
epoxides, 139, 141
epoxy, 140, 141, 142, 144, 169
epoxy groups, 142
EPR, 104, 106, 150, 151, 152, 156, 157
equilibrium, 19, 21, 25, 77, 80
ester, 140, 141, 142, 154
esterase, 5
esters, 141, 155
etching, 97, 165
ethanol, 172
ethers, 139
ethyl acetate, 183
ethylene, 138, 152, 155, 156
ethylene glycol, 152, 155, 156

evaporation, 169
evolution, 54, 55, 60, 70
EXAFS, 67, 68
excitation, 57, 58, 90, 105, 108, 194
experimental condition, 19
exposure, 138, 145, 146
extinction, 96, 116, 164
extraction, 15, 17, 18, 31
extrusion, 197

F

fabrication, 46, 170
failure, 106
FAS, 36
feeding, 54
fiber, 3, 7, 13, 14, 29, 31, 34, 35, 138, 200
fiber optics, 3
fibers, 34, 35, 137, 200
film, 6, 7, 9, 10, 12, 13, 35, 36, 37, 43, 45, 54, 63, 69, 97, 138, 142, 143, 144, 145, 163, 167, 168, 169, 171, 172, 174, 175, 176, 177
film formation, 35
films, 34, 36, 37, 45, 54, 63, 64, 69, 71, 137, 141, 163, 168, 169, 171, 174, 176, 177, 197
filtration, 75, 131
fixation, 132, 157
flow, 1, 3, 4, 6, 8, 9, 11, 13, 15, 16, 18, 20, 22, 24, 25, 27, 28, 29, 30, 96, 97, 98, 99
flow rate, 9, 12, 96, 98, 99
fluid, 30
fluorescence, 92, 93, 111, 112, 115, 116, 117, 158, 173, 181, 199
fluorination, 133
folate, 156
folic acid, 157
food, vii, 8, 166
food additives, 166
formamide, 76
fossil, vii, 86
fossil fuel, vii, 86
fossil fuels, vii, 86
free energy, 50, 76, 82, 107
free radical, 128
free volume, 39
friction, 198
frontal lobe, 8, 12
fuel, 50, 53, 75, 76, 77, 81, 82, 83, 85, 86
fuel cell, 50, 75, 76, 77, 81, 82, 83, 85, 86
functionalization, 152, 157
furan, 92
fusion, 157, 158, 159, 160

G

gas, 24, 25, 53, 56, 58, 60, 70, 76, 78, 83, 85, 86, 172
gas diffusion, 85
gas phase, 56, 58, 60
gases, 58, 78, 171
gasoline, 76, 86
gel, 36, 157
gene, 159
gene expression, 159
gene transfer, 159
General Electric, 85
generation, 45, 72, 81, 82, 84, 86, 90, 100, 115
generators, 141
genes, 158, 159, 160
genomics, 3
Gibbs, 50, 76
Gibbs free energy, 50, 76
glass, 1, 2, 15, 17, 34, 37, 39, 40, 41, 42, 43, 44, 45, 46, 63, 97, 164, 169, 174, 192
glass transition, 192
glass transition temperature, 192
glasses, 36, 40, 41, 42, 46
global warming, 85
glucose, 1, 2, 5, 6, 8, 11, 12, 13, 29, 30, 31, 50
glucose metabolism, 1, 11
glucose oxidase, 5, 31
glutamate, 2, 5, 6, 8, 9, 10, 11, 13, 29, 30
glutamic acid, 154, 155
glutaraldehyde, 5, 12, 26
glycerol, 152
glycoproteins, 3
goals, 1, 42
gold, 152, 163, 164, 165, 167, 168, 169, 171, 173, 174, 175, 176, 177

gold nanoparticles, 163, 164, 165, 167, 171
government, iv
GPC, 183
graduate students, viii
graphite, 183
green fluorescent protein, 159
greenhouse, vii, 76
greenhouse gas, 76
greenhouse gases, 76
groups, 26, 100, 115, 116, 118, 128, 142, 152, 153, 154, 157, 158, 181, 183, 185, 186, 188, 191, 192, 193, 195, 196
growth, vii, 35, 49, 156, 165
guidelines, 41

H

H_2, 49, 50, 52, 53, 54, 55, 70, 72, 76, 78, 81, 82, 84, 86, 167
halogenated, 133
halos, 115, 182
handling, 4, 173
health, 49
health problems, 49
heat, 15, 16, 36, 40, 42, 76, 77, 151, 152
heat capacity, 15
heat pumps, 77
heating, 16, 77, 82, 140, 146, 157, 192
heavy metal, 138, 164, 165
heavy metals, 138, 165
helix, 195
high pressure, 195
high temperature, 39, 80, 185, 200
HMS, 68, 69
HOMO, 56, 67, 112, 116, 117
homopolymers, 188
Honda, 50, 72
hormone, 3
hormones, 3
Horseradish peroxidase, 27
host, 64, 114
hot water, 36
human, 33
hybrid, 40, 79, 133, 159
hybridization, 174, 176
hydration, 157
hydride, 75, 77, 86, 87
hydrides, 79, 80
hydro, 27, 36, 50, 64, 86, 115, 139, 155, 156, 157, 172
hydrocarbon, 169
hydrocarbons, 50, 86, 115
hydrogen, 5, 6, 7, 13, 26, 50, 54, 75, 76, 77, 78, 80, 86, 87, 91, 126, 131, 185
hydrogen atoms, 185
hydrogen bonds, 76
hydrogen gas, 76, 78
hydrogen peroxide, 5, 6, 7, 13, 26
hydrolysis, 138
hydrolyzed, 36, 37
hydrophilicity, 12, 138
hydrophobic groups, 157
hydrophobic interactions, 156, 169
hydrophobicity, 138, 154, 172
hydrosilylation, 189
hydroxide, 24, 25, 76, 77
hydroxyl, 142
hydroxyl groups, 142
hydroxypropyl, 151

I

identification, 137
images, 175
imaging, 138
immersion, 80
immobilization, 3, 14, 18, 26
immunogenicity, 155
implementation, 170
impregnation, 60, 67
impurities, 9
IMS, 71
in situ, 22, 27, 29, 125
in vitro, 30
in vivo, 1, 2, 3, 4, 8, 9, 10, 12, 13, 28, 29, 30
India, 148
indium, 113
indium tin oxide (ITO), 113
industrial application, 199
industrial production, 200

industrialization, vii
industry, 49, 165, 171
infinite, 164
inflammation, 151
information technology, 178
infrared, 152
infrared light, 152
inhibition, 145
inhibitors, 3
injection, 1, 3, 4, 8, 9, 11, 13, 28, 30, 31, 112, 113, 156
injections, 12
injury, iv
innovation, 33, 160
inorganic, 40, 42, 45, 47
insertion, 125, 126, 139
insulators, 34, 197
integration, 118
interaction, 3, 68, 93, 104, 106, 108, 117, 151, 152, 168, 171
interactions, 67, 100, 150, 152, 153, 155, 156, 158, 169
interface, 1, 3, 15, 16, 17, 22, 44
interference, 12, 28
intermolecular, 130
internalization, 151, 160
intravenous, 30
inulin, 151
ion implantation, 65, 68
ion transport, 38, 40
ionization, 192
ions, 15, 18, 21, 34, 35, 38, 39, 41, 62, 63, 65, 66, 67, 68, 76, 78, 164, 172
Iron Age, 33
isomers, 130
ITO, 113, 114

J

Japan, 33, 80, 137
jewelry, 34

K

K^+, 32
ketones, 131
kidney, 150
kinetic energy, 62
kinetics, 106, 108

L

L2, 15
lactic acid, 155
lamina, 15, 16, 18, 99
landfills, 137
Langmuir, 161, 179
large-scale, 38, 81, 172
laser, 90, 105, 106, 107, 108
lattice, 67, 80
lattices, 77
law, 19, 25, 96
lectin, 4
LED, 200
lens, 18, 20
lifetime, 14, 58, 116, 117
ligand, 125, 150, 151
ligands, 60, 123, 124, 131, 151, 156, 157
light beam, 55
light scattering, 38
light-emitting diodes, 109, 112
limitations, 186
linear, 6, 7, 22, 138, 140, 142, 151, 152, 182, 183, 187, 194
linear polymers, 138, 140, 151
linkage, 142, 145, 146
lipids, 159
liposome, 156, 157, 158, 159
liposomes, 149, 156, 157, 158, 159
liquid chromatography, 2
liquid phase, 25, 54
liquids, 38
lithium, 40, 41, 42, 44, 45, 46, 47, 81, 102
liver, 156
living environment, 64
LOD, 12

London, 119, 177
long distance, 34, 67
low molecular weight, 13, 183, 185, 197
low temperatures, 80
low-temperature, 91, 109
luminescence, 114, 115, 134
LUMO, 56, 67, 112, 116, 117
lysine, 9, 151
lysosome, 159
lysosomes, 151, 154

M

M.O., 30
macromolecules, 150
magnesium, 183
magnet, 34
magnetic field, 151, 196, 199
magnetron, 54, 69, 70, 72
magnetron sputtering, 54, 69, 70, 72
maintenance, 79, 85
manganese, 79, 131
manufacturing, vii, 170
matrix, 40, 41, 107, 109, 110, 112, 114, 170, 171
MCP, 103, 104, 106, 107, 109, 112, 113
measurement, 6, 9, 12, 14, 20, 21, 30, 163
mechanical energy, 77, 82
media, 90, 165
melting, 192
melts, 41
membranes, 18, 22, 24, 86, 156, 157
mesoporous materials, 68
metabolic disorder, 2, 11
metabolism, 1, 11
metabolomics, 3
metal ions, 65, 67, 68
metal nanoparticles, 164
metals, 36, 80, 123, 182, 185, 188
methacrylic acid, 143
methane, 76, 86
methanol, 76, 83, 143, 146
methyl group, 193
methylcyclohexane, 110
methylene, 107
micelles, 149, 155, 156
microbes, 3
microchip, 4, 16, 18, 19
microdialysates, 30
microdialysis, 1, 2, 3, 5, 6, 7, 8, 9, 10, 11, 12, 13, 14, 28, 29, 30, 31
microelectronics, 139
microfluidic devices, 1, 2, 29
micrometer, 25, 96
microscope, 18
microscopy, 158
microstructure, 36, 44
microwave, 106
microwave radiation, 106
migration, 39, 172
military, 85
mineralization, 53
mineralized, 62
miniaturization, 173
mixing, 15, 27, 46, 85, 99
mobile phone, 79, 81
mobility, 198, 199
moieties, 133, 157, 185
molar ratio, 142, 167
molecular oxygen, 108, 114, 123, 131
molecular structure, 12
molecular weight, 13, 139, 150, 152, 183, 185, 197
molecular weight distribution, 152, 183
molecules, 15, 25, 38, 53, 58, 60, 76, 90, 94, 100, 103, 109, 112, 113, 115, 117, 133, 149, 150, 152, 153, 154, 155, 156, 157, 158, 164, 171, 173, 177
monomer, 22, 142, 145, 183, 185, 186
monomers, 22, 137, 138, 139, 140, 143, 144, 145, 146, 182, 183, 185, 187, 191
motion, 12, 34
movement, 8
MPS, 191
multifunctional monomers, 139
multiplicity, 111, 112, 113
muscle, 1, 29

N

Na^+, 32, 38
NaCl, 9, 22
NAD, 9
NADH, 9, 10
Nafion, 85, 86
nanoclusters, 164
nanocrystal, 36
nanocrystals, 36
nanodevices, 170
nanomaterials, 149, 160, 164
nanometer, 36, 150, 160
nanometer scale, 36
nanometers, 36, 150
nanoparticles, 85, 163, 164, 165, 168, 170, 174
nanostructures, 170
nanotechnology, 164
nanotubes, 164
nanowires, 164
naphthalene, 90, 91, 92, 93, 95, 97
natural, 49, 50, 81, 83, 86, 90
natural gas, 83, 86
Nd, 80
nerve, 29, 30
nerve cells, 29, 30
network, 46, 139, 140, 142, 147, 154
network polymers, 139
neurotransmitters, 1, 2, 8
New York, 73, 119, 120, 135, 148
next generation, 149
Ni, 65, 77, 78, 79, 80, 81
nickel, 75, 77, 86, 87, 165
nitrogen, 108, 134, 145, 146
NMR, 91, 134
non-crystalline, 36
normal, 27, 90, 149, 150
novel polymers, 140
n-type, 52
nuclear, 104
nuclei, 165, 192
nucleic acid, 3, 4
nucleophiles, 89, 102
nucleus, 159
nutrient, 2, 11
nylon, 22, 24, 25, 26, 27

O

observations, 71
oligomer, 140, 185
oligomeric products, 185
oligomers, 138, 183, 185
oligonucleotides, 158, 174
online, 1, 3, 5, 9, 11, 13, 28, 29, 30, 31
optical fiber, 34
optical properties, 171, 182, 192
optical systems, 173
optics, 3
optimization, 80
optoelectronic, 3
organelle, 4, 160
organic compounds, 50, 52, 53, 54, 62, 70, 76
organic peroxides, 131
organic polymers, 197
organic solvent, 76, 134
organic solvents, 76
oxidants, 52, 131, 181, 192
oxidation, 5, 50, 52, 54, 56, 62, 70, 71, 100, 106, 123, 131, 133, 165, 200
oxidative, 63, 90, 125, 126, 154
oxide, 41, 50, 51, 53, 56, 57, 58, 60, 64, 67, 69, 72, 83, 145, 155
oxides, 35, 58, 59, 60, 67, 68, 131
oxygen, 10, 50, 53, 62, 76, 78, 107, 108, 114, 123, 131, 145
oxygenation, 103, 107, 108
ozone, 49

P

paclitaxel, 156
PAFC, 83, 84
palladium, 125, 165
particles, 34, 38, 39, 43, 45, 52, 53, 56, 57, 64, 85, 150, 152, 164, 166, 168, 169, 171, 172, 177
passive, 71, 150, 151

pathways, 109
patterning, 2
Pb, 79
PCR, 16
PDMS, 97
PEMFC, 83
permeability, 6, 7, 12, 150
permeation, 22, 24, 25, 26, 27, 28, 29, 150
peroxide, 8, 14, 26, 28, 108
peroxide radical, 108
perturbation, 71
PET, 90, 100, 102, 103, 104, 105, 106, 107, 108, 109, 115
petroleum, 49
pH, 6, 7, 9, 12, 18, 20, 24, 151, 154, 157, 158, 159, 174
phenol, 140
phenolic resins, 141
phenolphthalein, 24, 25
phosphate, 6, 9, 12, 170
phospholipids, 157
phosphorescence, 117
phosphorus, 127
photocatalysis, 50, 51, 62
photocatalysts, 50, 53, 56, 57, 59, 60, 62, 63, 65, 66, 67, 69, 72, 86
photoconductivity, 181, 182, 197
photodegradation, 138
photoexcitation, 117
photographs, 17, 22
photoinduced electron transfer, 89, 90
photoirradiation, 97, 98, 99, 103, 115, 128
photolithography, 2
photoluminescence, 57, 58, 110, 111
photolysis, 53, 90, 105, 106, 107, 108
photon, 50, 52
photons, 116
photopolymerization, 138, 139
photosynthesis, 49, 50, 51, 54
physical properties, 138, 181, 192
physicochemical properties, 150, 164, 171
piezoelectric, 4
plants, 50, 51, 54
plasma, 69
plastic, 1, 8, 137, 163, 164, 165, 166, 167, 169, 170, 177
plastics, 36, 137, 138, 165
platinum, 6, 7, 9, 10, 13, 14, 126
play, 56, 133, 137, 195
PMS, 188, 189
poisons, 124
polarizability, 76
polarization, 82, 104, 106
pollutants, 82
pollution, vii, 49, 62, 165
poly(2-hydroxyethyl methacrylate), 146
poly(dimethylsiloxane), 97
poly(ethylene terephthalate), 138
polyamide, 22
polycarbonate, 138
polycondensation, 22, 138
polycrystalline, 35
polydispersity, 186
polyester, 138, 152
polyesters, 138
polyethylene, 171, 191
polymer, 6, 7, 14, 18, 24, 30, 49, 83, 134, 137, 138, 139, 140, 141, 142, 143, 144, 149, 150, 151, 152, 155, 156, 157, 158, 160, 181, 188, 190, 191
polymer chains, 157, 158
polymer film, 6, 144
polymer films, 144
polymer synthesis, 49, 138
polymer-based, 149, 150, 160
polymeric materials, 187
polymerization, 22, 126, 130, 134, 137, 138, 139, 146, 182, 183, 186, 187, 191
polymers, 134, 137, 138, 139, 140, 142, 143, 150, 151, 154, 155, 156, 157, 158, 181, 182, 191, 192, 197, 199
polymorphism, 163
polymorphisms, 174
polypeptide, 152
polysaccharides, 151
polystyrene, 113, 167, 168, 169
polyurethane, 138
poor, 44, 107, 113, 114, 130, 185, 187
porous materials, 35

potassium, 18, 20, 21, 32, 183
potential energy, 39, 195
powder, 35, 38, 42, 43, 85
powders, 34, 35, 42, 43, 46, 56
power, 38, 67, 69, 75, 79, 81, 82, 84, 85
power generation, 82, 84
precipitation, 165, 183
preference, 130
pressure, 27, 76, 77, 78, 80, 85, 86, 126, 131, 172, 195
printing, 138, 139, 147
probe, 6, 7, 8, 9, 10, 12, 14, 29, 174, 176, 177
production, vii, 49, 50, 51, 52, 53, 54, 72, 149, 160, 189, 199, 200
property, iv, 143, 151, 152, 154, 155, 157, 159, 172, 177
propylene, 155
prosperity, vii
protection, viii, 12, 51, 79, 133
proteins, 3, 7, 28, 151, 152, 158
proteomics, 3
protocols, 176, 177
proton exchange membrane, 54, 85
pseudo, 108
PSS, 113, 114
PTFE, 5, 6, 85
pulse, 106
pumping, 18, 20, 21
pumps, 18, 77
pure water, 75
purification, 34, 60, 62
PVP, 141
pyrene, 93, 115, 116, 117, 118
pyruvate, 2, 11, 12, 13, 30

Q

quantization, 56, 67
quantum, 101, 109, 112, 115, 116, 117, 164, 199
quantum dot, 164
quantum dots, 164
quantum yields, 116, 199
quartz, 37, 69, 70, 71, 96

R

radiation, 106, 117, 148
radical copolymerization, 143
radical polymerization, 142
radius, 195
rainwater, 75
Raman, 42
Raman spectroscopy, 42
random, 39, 157
range, 14, 34, 69, 181
rare earth, 80, 123, 124, 132
rare earth elements, 124
rare earths, 132
rat, 8, 9, 11, 12, 13, 28, 30
rats, 29
raw material, 33, 34, 35, 81
raw materials, 33, 34, 35
reactants, 52, 114
reaction mechanism, 60, 90
reaction temperature, 90, 91, 183, 185
reactive groups, 152
reactive sites, 140
reactivity, 12, 53, 56, 59, 60, 62, 67, 72, 99, 104, 108, 128
reagent, 3, 18, 20, 138
reagents, 21, 124, 131, 133
receptors, 3, 150
recognition, 2, 3, 151
recombination, 52, 53, 67, 109
recovery, 154
recycling, 9, 10, 30, 31, 137, 138
red shift, 68, 196
redistribution, 187
redox, 52, 101, 133
redox-active, 133
reflection, 36
refractive index, 118
regeneration, 104, 111, 112, 113, 125
regioselectivity, 92, 97, 100, 125, 128
regular, 80
relationship, 59, 60
reliability, 42, 47
remediation, 53, 65, 72
repair, 139

research and development, 85
reservoir, 76
residues, 154
resin, 97, 170
resins, 146
resistance, 47, 163, 168, 170, 171, 174, 177
resistive, 200
resistivity, 168
response time, 1, 19, 25
retention, 150, 154
Reynolds, 16, 99
Reynolds number, 16, 99
RFA, 31
rhodium, 126
rings, 91
robustness, 169
room temperature, 40, 41, 42, 60, 76, 80, 106, 146, 200
room-temperature, 42
ROP, 182, 187
roughness, 36, 38
rubber, 137

S

safety, 42, 47, 79, 81
salt, 138
salts, 84, 165, 187
sample, 1, 3, 6, 9, 10, 11, 18, 20, 146, 173, 174, 176
sampling, 1, 2, 3, 9, 11, 13, 28, 29, 30, 31
scatter, 36
scattering, 34, 38
scavenger, 52
search, 56
secrets, 165
selecting, 171, 176
selectivity, 3, 18, 59, 60, 91, 99, 124, 135, 177, 195
selenium, 124, 125, 126, 127
Self, 63, 79
self-assembly, 173, 175
SEM, 36, 167, 168, 175
semiconductor, 52, 86
semiconductors, 50, 53, 65, 194
sensing, 2, 3, 12, 17, 18, 19, 20, 29, 30, 32, 173, 174
sensitivity, 1, 2, 8, 14, 154, 177
sensor technology, 17, 29
sensors, 3, 4, 14, 18, 29, 32, 77, 164, 171, 172
separation, 53, 109, 175
serum, 7, 12, 32
services, iv
shape, 34, 152
shaping, 35
shoulder, 65
side effects, 149
signals, 3, 11, 20, 106
silane, viii
silica, 7, 14, 34
silica glass, 34
silicon, 101, 117, 124, 181, 182, 183, 188, 192, 193, 195, 200
silver, 13, 14
silylenes, 197
single nucleotide polymorphism, 163, 174
sintering, 35
SiO2, 45
sites, 1, 61, 140, 149, 150, 151, 156, 157, 185
SNP, 174, 177
sodium, 6, 18, 21, 24, 25, 26, 38, 138, 183, 184
sodium hydroxide, 24, 25
SOFC, 83, 84
solar, 49, 50, 53, 54, 55, 56, 65, 67, 70, 72, 81, 86
solar cell, 86
solar cells, 86
solar energy, 49, 50, 54, 56, 67, 72
sol-gel, 36, 199
solid-state, 34, 42, 43, 44, 45, 46, 47, 195, 197
solubility, 76, 138, 139
solvent, 29, 85, 105, 107, 110, 118, 131, 133, 183, 185
solvents, 90, 100, 123, 124, 133, 139, 140, 142, 143, 146, 185, 196
species, 12, 19, 24, 25, 26, 27, 28, 56, 57, 58, 60, 67, 72, 90, 100, 102, 109, 110, 112, 113, 114, 115, 125, 128, 132, 138, 191, 196, 197

specific surface, 35
specificity, 9, 150, 156, 157
spectroscopy, 104
spectrum, 57, 58, 106, 117, 134, 200
speed, 9
spin, 34, 97, 112
spleen, 156
sputtering, 54, 69
stability, 3, 53, 85, 86, 138, 155, 164, 182
stabilize, 157
stages, 10
stainless steel, 7
steel, 7
Stimuli, 154
STM, 174
Stokes shift, 117
storage, 38, 50, 76, 77, 80, 81, 86
strategies, 112, 150
streams, 17
strength, 152, 153, 200
students, viii
styrene, 138, 151, 156
subcutaneous tissue, 1, 2, 29, 30
substances, vii, 75
substitution, 116, 124, 188
substitution reaction, 124, 188
substrates, 2, 3, 15, 27, 35, 36, 37, 99, 126, 133, 139, 164, 169
Sulfide, 40, 42, 45
sulfur, 38, 45, 125, 126
sulfuric acid, 76
Sun, 30
sunlight, 54, 55, 123, 124, 131, 132
superlattice, 87
superlattices, 164
superposition, 106
supply, 82, 185
suppression, 80
surface layer, 165
surface modification, 15, 22, 26, 27, 80, 156, 157, 158
surface structure, 36
switching, 18
symbols, 137
synchrotron, 42
Synergy, 157
synthesis, vii, 42, 49, 90, 95, 102, 123, 124, 131, 133, 135, 138
synthetic polymers, 137, 151

T

tanks, 86
targets, 151
Teflon, 36
TEM, 71
temperature, 16, 39, 40, 42, 54, 69, 83, 91, 107, 109, 112, 114, 140, 146, 151, 154, 155, 156, 157, 158, 183, 185, 193, 195
temperature dependence, 40
tensile, 200
tensile strength, 200
terminals, 154, 157, 170, 172, 173
tetrachloroethylene, 62, 172
therapy, 152
thermal activation, 109
thermal degradation, 146, 147
thermal energy, 82
thermal lens, 18, 20
thermal properties, 139, 147
thermal treatment, 141, 198
thermoluminescence, 89, 109, 118
thermolysis, 108, 143
thin film, 36, 37, 54, 55, 63, 64, 69, 70, 71, 72, 146
thin films, 36, 54, 63, 64, 69, 70, 71
tin, 46, 113, 165
tin oxide, 113
TiO_2, 50, 52, 53, 54, 55, 56, 57, 59, 62, 63, 64, 65, 66, 67, 69, 70, 71, 72
tissue, 1, 2, 3, 4, 29, 30
titanium, 50, 51, 53, 56, 57, 58, 59, 60, 67, 69, 72
Titanium oxide, 67
Tokyo, 47
tolerance, 85
toluene, 172, 183
topology, 151
toxic substances, vii
toys, 79

TPO, 145
trans, 91
transcription, 159
transducer, 3
transfection, 159
transfer, 15, 34, 47, 57, 58, 60, 68, 78, 93, 100, 104, 115, 158, 159, 166, 168, 174, 182, 183, 191, 192
transferrin, 157
transformation, 77
transformations, 123, 127
transistors, 4
transition, 40, 65, 68, 112, 118, 123, 124, 126, 127, 128, 131, 132, 133, 157, 185, 194
transition metal, 68, 124, 126, 127, 128, 131, 185
transition metal ions, 68
transition temperature, 40, 192
transitions, 113
transmission, 34, 69
transmittance spectra, 37
transparency, 34, 38, 97
transparent, 34, 36
transport, 24, 27, 38, 39, 40, 113, 194, 197, 199
transportation, 16
triflic acid, 146
triphenylphosphine, 125
tubular, 30
tumor, 150, 151, 156, 157, 158
tumor cells, 151, 157, 158
tumor growth, 156
tungsten, 97, 134
tunneling, 168, 172
turbulent, 17, 99

U

ultrasound, 151
ultraviolet, 53, 138, 139, 181
ultraviolet light, 139
unsaturated polyesters, 138
UV, 50, 52, 53, 56, 58, 60, 62, 63, 64, 65, 66, 67, 68, 69, 70, 71, 97, 116, 128, 130, 134, 138, 139, 140, 141, 142, 143, 144, 145, 147, 148, 182, 193, 199
UV absorption, 97
UV absorption spectra, 97
UV irradiation, 60, 141, 142, 143, 144
UV light, 52, 53, 56, 58, 60, 62, 63, 64, 67, 69, 97, 130, 143, 199

V

vacuum, 53
valence, 52, 56
values, 96, 106, 113, 118, 168, 174, 194, 197
van der Waals, 195
vanadium, 68, 131
vapor, 76, 172
variation, 20
vehicles, 40, 79, 85, 86, 154
vessels, 96, 150
vibration, 34, 107
vinyl monomers, 138
viscosity, 138
visible, 36, 37, 50, 51, 53, 54, 55, 62, 63, 65, 66, 67, 69, 71, 72, 124, 128, 130, 132, 137, 152
vitrification, 34, 35, 39

W

waste disposal, 165
wastes, 166
wastewater, 75
water, vii, 8, 36, 37, 49, 50, 52, 53, 54, 62, 63, 64, 72, 75, 76, 77, 78, 81, 82, 86, 123, 132, 143, 144, 154, 156, 166, 172
water vapor, 76
water-soluble, 143
wavelengths, 53, 65, 67, 69, 116, 128, 195
wettability, 64
wind, 81

X

XANES, 67

X-ray diffraction, 42

Y

yield, 100, 101, 102, 103, 107, 112, 117, 151, 183, 185, 191, 197

Z

zeolites, 50, 57, 61, 67
zirconia, 38, 84
Z-isomer, 130
ZnO, 52